Thinking Ahead—Essays on Big Data, Digital Revolution, and Participatory Market Society

Dirk Helbing

Thinking Ahead—Essays on Big Data, Digital Revolution, and Participatory Market Society

 Springer

Dirk Helbing
ETH Zürich, CLU E1
Computational Social Science
Claussiusstrasse 50
8092 Zürich
Switzerland

ISBN 978-3-319-15077-2   ISBN 978-3-319-15078-9 (eBook)
DOI 10.1007/978-3-319-15078-9

Library of Congress Control Number: 2015934446

Springer Cham Heidelberg New York Dordrecht London

Cover illustration: Internet security and privacy issues with a human eye and digital binary code as surveillance of hackers or hacking from cyber criminals watching prohibited access to web sites with firewalls. Stock Photo. Copyright: Lightspring.

Springer is part of Springer Science+Business Media (www.springer.com)

About the Author

Physicist Dirk Helbing is Professor of Computational Social Science at the Department of Humanities, Social and Political Sciences and an affiliate of the Computer Science Department at ETH Zurich, as well as co-founder of ETH's Risk Center. He is internationally known for the scientific coordination of the FuturICT Initiative which focuses on using smart data to understand techno-socio-economic systems.

"Prof. Helbing has produced an insightful and important set of essays on the ways in which big data and complexity science are changing our understanding of ourselves and our society, and potentially allowing us to manage our societies much better than we are currently able to do. Of special note are the essays that touch on the promises of big data along with the dangers...this is material that we should all become familiar with!"

Alex Pentland, MIT, author of Social Physics: How Good Ideas Spread—The Lessons From a New Science

"Dirk Helbing has established his reputation as one of the leading scientific thinkers on the dramatic impacts of the digital revolution on our society and economy. Thinking Ahead is a most stimulating and provocative set of essays which deserves a wide audience."

Paul Ormerod, economist, and author of Butterfly Economics and Why Most Things Fail.

"It is becoming increasingly clear that many of our institutions and social structures are in a bad way and urgently need fixing. Financial crises, international conflicts, civil wars and terrorism, inaction on climate change, problems of poverty, widening economic inequality, health epidemics, pollution and threats to digital privacy and identity are just some of the major challenges that we confront in the twenty-first century. These issues demand new and bold thinking, and that is what Dirk Helbing offers in this collection of essays. If even a fraction of these ideas pay off, the consequences for global governance could be significant. So this is a must-read book for anyone concerned about the future."

Philip Ball, science writer and author of Critical Mass

"This collection of papers, brought together by Dirk Helbing, is both timely and topical. It raises concerns about Big Data, which are truly frightening and disconcerting, that we do need to be aware of; while at the same time offering some hope that the technology, which has created the previously unthought-of dangers to our privacy, safety and democracy can be the means to address these dangers by enabling social, economic and political participation and coordination, not possible in the past. It makes for compelling reading and I hope for timely action."

Eve Mitleton-Kelly, LSE, author of Corporate Governance and Complexity Theory and editor of Co-evolution of Intelligent Socio-technical Systems.

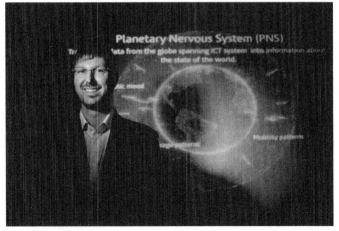

Foto: Sabina Bobst

Preface

This booklet presents a collection of essays and discussion or white papers on Big Data, the ongoing Digital Revolution and the emergent Participatory Market Society. These have been written since the year 2008 in anticipation of and response to the financial and other crises. While we have seen a pretty peaceful period after the fall of the Berlin Wall in 1989, the world seems to have increasingly destabilized in the aftermath of September 11, 2001.

If we want to master the related challenges, we must analyze the underlying problems and change the way we manage our techno-socio-economic systems.

I would like to thank many friends and colleagues, in particular the worldwide FuturICT community, for the inspiring discussions and the continued support. I am also grateful to Stefano Balietti, James Breiding, and Markus Christen for their reprint permissions regarding two of the chapters in this booklet.

November 2014   Dirk Helbing
Zürich

Contents

Contents

1

Introduction—Have We Opened Pandora's Box?

This chapter first appeared in the FuturICT Blog on September 10, 2014, see http://futurict.blogspot.ch/2014/09/have-we-opened-pandoras-box_10.html, and is reproduced here with minor stylistic improvements. Acknowledgments: I would like to thank many friends and colleagues, in particular the world-wide FuturICT community, for the inspiring discussions and the continued support. I am also grateful to Stefano Balietti, James Breiding, and Markus Christen for their reprint permissions regarding two of the chapters in this booklet.

1.1 Global Financial, Economic and Public Spending Crisis

The first of the contributions in this booklet dates back to March 2008, when Markus Christen, James Breiding and myself became concerned about the stability of the financial system that we felt urged to write a newspaper article to alert the public (see the English translation in Chap. 4). Unfortunately, at that time, the public was not ready to listen. Newspaper editors found our analysis too complex. We responded that a financial crisis would be impossible to prevent, if newspapers failed to explain the complexity of problems like this to their audience. Just a few months later, Lehmann Brothers collapsed, which gave rise to a large-scale crisis. It made me think about the root causes of economic problems [1–4] and of global crises in general [5, 6] (see Chaps. 4, 5, and 7). But my collaborators and I saw not only the financial crisis coming. We also voiced the surveillance problem early on and the political vulnerability of European gas supply. We studied conflict in Israel, the spreading of diseases, and new response strategies to earthquakes and other disasters. Shortly after, all of this turned out to be highly relevant, almost visionary.

When I attended a Global Science Forum in 2008 organized by the OECD [7], most people still expected that the problems in the US real estate market and the banking system could be fixed. However, it was already clear to me and probably also to many other complexity scientists that they would cause cascade effects and trigger a global economic and public spending crisis, which we would not recover from for many years. At that time, I said that nobody understood our financial system, our economy, and our society well enough to grasp the related problems and to manage them successfully. Therefore, I proposed to invest into a large-scale project in the social sciences—including economics—in the very same way as we have invested billions into the CERN elementary particle accelerator, the ITER fusion reactor, the GALILEO satellite system, space missions, astrophysics, the human genome projects, and more. I stressed that, in the twenty-first century, we would require a "knowledge accelerator" to keep up with the pace at which our societies are faced with emerging problems [8]. Today, business and politics are often based on scientific findings that are 30 to 50 year old, or not based on evidence at all. This is not sufficient anymore to succeed in a quickly changing world. We would need a kind of Apollo project, but not one to explore our universe—rather one that would focus on the Earth and what was going on there, and why.

1.2 Need of a "Knowledge Accelerator"

As a consequence, the VISIONEER support action funded by the European Commission (http://www.visioneer.ethz.ch) worked out four white papers proposing large-scale data mining, social supercomputing, and the creation of an innovation accelerator [9]. Already back in 2011, VISIONEER was also pointing out the privacy issues of modern information and communication technologies, and it even made recommendations how to address them [10].

Then, in response to the European call for two 10-year-long one billion EURO flagship projects in the area of Future Emerging Technologies (FET), the multi-disciplinary FUTURICT consortium was formed to turn this vision into reality (see http://www.futurict.eu). Thousands of researchers world-wide, hundreds of universities, and hundreds of companies signed up for this. 90 million € matching funds were waiting to be spent in the first 30 months. But while the project was doing impressively well, to everyone's surprise it was finally not funded, even though we proposed an approach aiming at ethical information and communication technologies [10, 11], with a focus on privacy and citizen participation [12].

This possibly meant that governments had decided against FuturICT's open, transparent, participatory, and privacy-respecting approach, and that they might invest in secret projects instead. If this were the case, a worrying digital arms race would result. Therefore, while spending my Easter holidays 2012 in Sevilla, I wrote a word of warning with the article "Google as God?" (see Chap. 9). Shortly later, Edward Snowden's revelations of global mass surveillance shocked the world, including myself [13]. These unveiled past and current practices of secret services in various countries and criticized them as illegal. Even though an informed reader could have expected a lot of what was then reported, much of it just surpassed the limits of imagination.

The sheer extent of mass surveillance, the lack of any limits to the technical tools developed, and the way they were used frightened and alarmed many citizens and politicians. The German president, Joachim Gauck, for example, commented: "This affair [of mass surveillance] concerns me a lot. ... The worry that our phone calls and emails would be recorded and stored by a foreign secret service hampers the feeling of freedom—and with this there is a danger that freedom itself will be damaged." [14] Nevertheless, many important questions have still not been asked: How did we get into this system of mass surveillance? What was driving these developments? Where will they lead us? And what if such powerful information and communcation technologies were misused? Such questions will be addressed in this booklet.

1.3   We are Experiencing a Digital Revolution

One of the important insights is: We are in the middle of a digital revolution—a third industrial revolution after the one turning agricultural societies into industrial ones, and these into service societies. This will fundamentally transform our economy and lead us into the "digital society" [15]. I claim that not only the citizens haven't noticed this process early enough, but also most businesses and politicians. By the time we got a vague glimpse of what might be our future, it had already pervaded our society, in the same way as the financial crisis had infected most parts of our economy. Again, we have difficulties to identify the responsible people—we are facing a systemic issue.

Rather than blaming companies or people, my effort is to raise awareness for the implications of the techno-socio-economic systems we have created: intended and unintended, positive and negative ones, and to point the way to a brighter future. As it turns out, we do in fact have better alternatives. But before I discuss these, let me first give a reasonably short summary of the current insights into the side effects of information and communication technologies, as far as they must concern us.

1.4   Threats to the Average Citizen

Let me begin with the implications of mass surveillance for citizens. It is naive and just wrong to assume mass surveillance would not matter for an average citizen, who is not engaged in any criminal or terrorist activities. The number of people on lists of terror suspects comprises a million names [16]—other sources even say a multiple of this. It became known that these lists contains a majority of people who are not terrorists nor linked with any. Furthermore, since friends of friends of contact persons of suspects are also under surveillance, basically everyone is under surveillance [17].

Of course nobody would argue against preventing terrorism. However, mass surveillance [18] and surveillance cameras [19] haven't been significantly more effective in preventing crime and terror than classical investigation methods and security measures, but they have various side effects. For example, tens of thousands of innocent subjects had to undergo extended investigation procedures at airports [20]. In connection with the war on drugs, there have even been 45 million arrests [21], where many appear to be based on illegal clues from surveillance [22]. Nevertheless, the war on drugs has failed, and US Attorney General Eric Holder finally concluded: "Too many Americans go to too many prisons for far too long, and for no truly good law enforcement reason" [23].

Recently, many people have also been chased for tax evasion. While I am not trying to defend drug misuse or tax evasion, we certainly see a concerning transition from the principle of assumed innocence to a situation where everyone is considered to be a potential suspect [24]. This is undermining fundamental principles of our legal system, and implies threats for everyone. In an over-regulated society, it is unlikely that there is anybody who would not violate any laws over the time period of a year [25]. So, everyone is guilty, in principle. People (and companies) are even increasingly getting into trouble for activities, which are legal—but socially undesirable, i.e. we are increasingly seeing phenomena comparable to "witch hunting." For example, in December 2013, thousands of people got sued by a law firm for watching porn [26]. For the first time, many people became aware that all of their clicks in the Internet were recorded by companies, and that their behavior was tracked in detail.

1.5 Threats so Big that One Cannot Even Talk About Them

On the side of the state, such tracking is being justified by the desire to prevent danger to society, and child pornography is often given as one of the reaons. Again, nobody would argue against the need to protect children from misuse, but this time the subject is even so taboo that most people are not even aware of what exactly one is talking about. You can't really risk to look up information in the Internet, and you are advised to delete photographs depicting yourself when you were a child. Only recently, we have learned that Internet companies report thousands of suspects of child pornography [27]. It is not known what percentage of these people have ever touched a child in an immoral way, or paid money for unethical pictures or video materials. This is particularly problematic, as millions of private computers are hacked and used to send spam mails [28]; illegal material might easily be among them.

Note that passwords of more than a billion email accounts have been illegally collected, recently [29]. This might imply that almost everyone living in a first world country can be turned into a criminal by putting illegal materials on one of their digital devices. In other words, if you stand in somebody's way, he or she might now be able to send you to prison, even if you have done nothing wrong. The evidence against you can be easily prepared. Therefore, your computer and your mobile device become really dangerous for you. It is no wonder that two thirds of all Germans don't trust that Internet companies and public authorities use their personal data in proper ways only; half of all Germans even feel threatened by the Internet [30].

1.6 Are we Entering an Age of Discrimination?

On the side of big business, our clicks are being collected for the sake of personalized advertisements, but also to make cash in ways that are highly problematic. Whenever you apply for a loan or a health insurance, between 3000 and 5000 personal data about you might be used to determine the rate you have to pay—or whether you get an offer at all. You would probably be shocked to see what is known about you, and how many thousands or millions of people in the world have access to these data. The people looking into our sensitive personal data, including health data, range from secret services over border police to banks and insurance companies to the businesses that sell and provide advertisements.

While these data are collected even if you don't implicitly agree to share them (by accepting the terms of use of a software, browser, or app), it has become common to apply them in increasingly more business areas. Some of the data that were collected without informed consent may be "whitewashed" by intermediary data providers buying illegal data and declaring their legal origin ("data laundry"). Personal data are used, for example, to make personalized offers when buying products on the Internet. In fact, offered products and prices now often depend on your country, neighborhood, and salary. In other words, if you live in the "wrong neighborhood," you may have to pay a higher price, and if you don't walk enough or if you frequently eat at fastfood restaurants, your life insurance may be more expensive. In other words, discrimination will increasingly become an issue (see Chap. 11). Besides, about half of the personal data sets contain mistakes [31]. As a consequence, you will get wrong offers without any chance to check and challenge them. It is obvious that we have currently a great lack of transparency, and also of mechanisms to get wrong data corrected.

1.7   Threats to Companies

But the age of Big Data is not only becoming a threat to citizens. The same applies to companies as well. There is an increasing risk of espionage of sensitive trade secrets. For example, it has been proven that intellectual property of the Enercon company was stolen and patented by a competing company [32]. One may wonder, how such cyber espionage works, but most computer systems are more vulnerable than one would think [33]. Every hour, there are thousands of cyberattacks, and it is often just a matter of time until one of them succeeds. It does not have to be a secretary who opened a virus or trojan horse attachment of an email. Stuxnet, for example, was even able to reach computers that are not directly connected to the Internet [34]. Any USB port may be a problem [35], too, and even your water boiler in the kitchen [36], not to talk about hardware backdoors [37, 38] or software vulnerabilities such as zero day exploits, which may spread by autoupdates [39].

Most mobile devices can be easily hacked [40], and the xKeyscore program is said to be able to follow the correspondence of any known e-mail address, and even keyboard entries as you type [41] (which also means that there is probably no safe password). As there are about a million people who have (had) access to data on the same level as Edward Snowden [42], among them mostly people working for private companies, business espionage may not necessarily involve government employees. It could as well be done by employees of private businesses with privileged access to the information infrastructure of a secret service or just similar technology. Encryption is only a partial protection. Many encryption methods have been weakened [43], not to talk about problems such as the heartbleed bug [44] or interfaces for remote service access [45]. This has made sensitive data, such as money transfers or health data highly vulnerable. The providers face increasing difficulties to guarantee data security. Many companies recently had to report the theft of sensitive data, and the same applies to public authorities, including the military [46].

1.8 Political and Societal Risks

However, the age of Big Data also implies considerable political and societal risks. The most evident threat is probably that of cyberwar, which seriously endangers the functionality of critical infrastructures and services. This creates risks that may materialize within milliseconds, for extended time periods, and potentially for large regions [47]. Therefore, the nuclear response to cyberattacks is considered to be an option [48]. Some countries also work on automated programs for responsive cyberattacks [49]. However,

as cyberattacks are often arranged such that they appear to originate from a different country, this could easily lead to responsive
counterstrikes on the wrong country—a country that has not been
the aggressor.

But there are further dangers. For example, political careers
become more vulnerable to what politicians have said or done
many years back—it can all be easily reconstructed. Such problems do not require that these people have violated any laws—it
might just be that the social norms have changed in the meantime. This makes it difficult for personalities—typically people
with non-average characters—to make a political career. Therefore, intellectual leadership, pointing the way into a different,
better future, might become less likely.

At the same time, Big Data analytics is being used for personalized election campaigns [50], which might determine a voter's
political inclination and undermine the fundamental democratic
principle of voting secrecy. With lots of personal and social media
data it also becomes easier to give a speech saying exactly what
the people in a particular city would like to hear—but this, of
course, does not mean the promises will be kept. Moreover, if
the governing political leaders have privileged access to data, this
can undermine a healthy balance of power between competing
political parties.

1.9   Are the Secret Services Democratically well Controlled?

It has further become known that secret services, also in democratic countries, manipulate discussions in social media and
Internet contents, including evidence, by so-called "cyber magicians" [51]. In South Korea, the prime minister is even said
to have been tweeted into office by the secret services [52]. But

not always are secret services playing in accord with the ruling politicians. In Luxembourg, for example, it seems they have arranged terror attacks (besides other crimes) to get a higher budget approved [53]. They have further spied on Luxembourg's prime minister Jean-Claude Juncker, who lost control over the affair and even his office. One may therefore hope that, in his current role as the president of the European Commission, he will be able to establish proper democratic control of the activities of secret services.

In fact, there is a serious but realistic danger that criminals might gain control of the powers of secret services, who should be protecting the society from organized crime. Of course, criminals will always be attracted by Big Data and cyber powers to use them in their interest, and they will often find ways to do so. In Bulgaria, for example, a politician is said to have been trying to gain control over the country's secret services for criminal business. The Bulgarian people have been demonstrating for many weeks to prevent this from happening [54].

1.10 What Kind of Society are we Heading to?

Unfortunately, one must conclude that mass surveillance and Big Data haven't increased societal, econonomic, and cyber security. They have made us ever more vulnerable. We, therefore, find our societies on a slippery slope. Democracies could easily turn into totalitarian kinds of societies,[1] or at least "democratorships," i.e. societies in which politicians are still voted for, but in which the

---

[1] Just imagine the mass surveillance data would get in control of extreme political parties, and they would use these to terrorize the people. Germany has had a Nazi and a Stasi regime, and knows, how terrible this can end.

citizens have no significant influence anymore on the course of events and state of affairs. The best examples for this are probably the secret negotiations the ACTA and TTIP agreements, i.e. laws intended to protect intellectual property and to promote free trade regimes. These include parallel legal mechanisms and court systems, which would take intransparent decisions that the public would nevertheless have to pay for.

It seems that traditional democracies are more and more transformed into something else. This would perhaps be ok, if it happened through an open and participatory debate that takes the citizens and all relevant stakeholders on board. In history, societies have undergone transformations many times, and I believe the digital revolution will lead us to another one. But if politicians or business leaders acted as revolutionaries trying to undermine our constitutional rights, this would sooner or later fail. Remember that the constitution—at least in many European countries—explicitly demands from everyone to protect privacy and family life, to respect the secrecy of non-public information exchange, to protect us from misuse of personal data, and to grant the possibility of informational self-determination, as these are essential functional preconditions of free, human, and livable democracies [55]. The secret services should be protecting us from those who question our constitutional rights and don't respect them. Given the state of affairs, this would probably require something like an autoimmune response. It often seems that not even public media can protect our constitutional rights efficiently. This is perhaps because they are not able to reveal issues that governments share with them exclusively under mutually agreed confidence.

1.11 "Big Governments" Fueled by "Big Data"

In the past years, some elites have increasingly become excited about the Singaporian "big government" model, i.e. something like an authoritarian democracy ruled according to the principle of a benevolent, "wise king," empowed by Big Data [56]. As logical as it may sound, such a concept may be beneficial up to a certain degree of complexity of a society, but beyond this point it limits the cultural and societal evolution (see Chaps. 9 and 13). While the approach to take decisions like a "wise king" might help to advance Singapore and a number of other countries for some time, in a country like Germany or Switzerland, which gain their power and success by engaging into balanced and fair solutions in a diverse and well-educated society with a high degree of civic participation, it would be a step backwards. Diversity and complexity are a precondition for innovation, societal resilience, and socio-economic well-being [15]. However, we can benefit from complexity and diversity only if we allow for distributed management and self-regulating systems. This requires to restrict top-down control to what cannot be managed in a bottom-up way. That is, where the transformative potential of information and communication systems really is: information technology can now enable the social, economic and political participation and coordination that was just impossible to organize before.

It is now the time for a societal dialogue about the path that the emerging digital society should take: either one that is authoritarian and top-down, or one that is based on freedom, creativity, innovation, and participation, enabling bottom-up engagement [57]. I personally believe a participatory market society is offering the better perspectives for industrialized services societies in the future, and that it will be superior to an authoritarian top-down approach. Unfortunately, it seems we are heading towards the latter. But it is important to recognize that the dangers of the current Big Data approach are substantial, and that there is nobody who could not become a victim of it. It is crucial to understand and admit that we need a better approach, and that it was a mistake to engage into the current one.

1.12   We Must Move Beyond September 11

The present Big Data approach seems to be one of the many consequences of September 11, which did not change our world to the better. By now, it has become clear that the "war on X" approach—where "X" stands for drugs, terror, or other countries—does not work. Feedback, cascade and side effects have produced many unintended results. In the meantime, one is trying to find "medicines" against the side effects of the medicines that were applied to the world before.

The outcome of the wars on Iraq and Afghanistan can hardly be celebrated as success. We rather see that these wars have destabilized entire regions. We are faced with increased terrorism by people considering themselves as freedom fighters, a chaotic aftermath of Arab spring revolutions, devastating wars in Syria, Israel and elsewhere, an invasion of religious warriors, increased unwelcomed migration, poverty-related spreading of dangerous diseases, and larger-than-ever public spending deficits; torture, Guantanamo, secret prisons, drones and an aggressive cybersecurity approach have not managed to make the world a safer place [58]. As these problems demonstrate, globalization means that problems in other parts of the world will sooner or later affect us [59].

In other words, in the future we must make sure that, if we want to have a better and peaceful life, others around us will also need to find peaceful and reasonable living conditions. To better understand the often unexpected and undesirable feedback, cascade and side effects occurring in the complex interdependent systems of our globalized world, it is important to develop a Global Systems Science [5]. For example, it has been recently pointed out, even from unexpected sides such as Standard and Poor's, that too much inequality endangers economic and societal progress [60]. It is also important to recognize that respect and qualified trust are a more sustainable basis for socio-economic order than power and fear [61] (see Chap. 10). I believe the dangerous aspect of mass surveillance is that its impact will become obvious only over a time period of many years. By the time we notice this, it might be too late to protect us from harm. Like nuclear radiation, one cannot directly feel the effects of mass surveillance, but it nevertheless causes structural damages—in this case to democratic societies. Mass surveillance undermines trust and legitimacy. However, trust and legitimacy are the fabric that keeps societies together—they create the power of our political representatives and public institutions. Without trust, a society becomes unstable.

1.13 What Needs to be Done

It is not unreasonable to be afraid of the "genie out of the bottle" that mass surveillance released. Some people consider it to be one of the things that escaped from Pandora's Box in the aftermath of September 11. But hope never dies. What can we do? First of all, to ensure accountability, it seems necessary to record each access to personal data (including the computational operation and the exact data set it was executed on). Second, one must restore lost trust by the public, which requires a sufficient level of transparency. For example, the log files of data queries executed by secret services and other public authorities would have to be accessible to independent and sufficiently empowered supervising authorities.

Similarly, log files of data queries executed by companies should be regularly checked by independent experts such as qualified scientists or citizen scientists. To be able to trust Big Data analytics, the public must know that it is scientifically sound and compliant with the values of our society and constitution. This also requires that users, customers, and citizens have a right to legally challenge results of Big Data analytics. For this, Big Data analytics must be made reproducible, such that the quality and law compliance of data mining results can be checked by independent experts.

Furthermore, it should be ensured that the power of Big Data is not used against the legitimate interests of people. For example, I recommend to use it to enable people, scientists, companies and politicians to take better decisions and more effective actions rather than applying it for the sake of large-scale law enforcement. The use of Big Data for criminal investigation should, therefore, be restricted to activites that endanger the foundations of a well-functioning society. It might further be necessary to punish data manipulation and data pollution, no matter who engages in it (including secret services).

Given the many instances of data manipulation today, data traces should not be considered as pieces of evidence themselves. Furthermore, for the sake of just and legitimate sanctioning systems, it must be ensured that sanctions are not applied in an arbitrary and selective way. In addition, the number of criminal investigations triggered by data analytics must be kept low and controlled by the parliament. Otherwise, in an over-regulated society, Big Data analytics might be misused by the elites to shape the society according to their taste—and this would surely end in a disaster sooner or later. In particular, the use of Big Data should not get into the way of freedom and innovation, as these are important functional success principles of complex societies.

It is also important to recognize that the emergent digital society will require particular institutions, as it was also the case for the industrial and the service societies. This includes data infrastructures implementing a "new deal on data" [62], which would give users control over their own data and allow them to benefit from profits created with them. This can be done with the "Personal Data Purse" approach, which has recently been developed to comply with the constitutional right of informational self-determination [63]. Further infrastructures and institutions needed by the digital society will be addressed in Chap. 13.

1.14   A Better Future, Based on Self-Regulation

Finally, I recommend to engage into the creation of self-regulating systems. These can be enabled by real-time measurements, which the sensor networks underlying the emerging "Internet of Things" will increasingly allow. Interestingly, such applications can support socio-economic coordination and order based on self-organization, without requiring the storage of personal or other sensitive data. In other words, the production of data and their use for self-regulating systems would be temporary and local, thereby enabling efficient and desirable socio-economic outcomes while avoiding dystopian surveillance scenarios. I am convinced that this is the information-based way into a better future and, therefore, I will describe further details of this approach in an upcoming book on the self-regulating digital society [64]. While the booklet in your hands is more focused on concerns related to the current trends and developments, the forthcoming book will be focused on the question, what we can do to promote a "happy end."

References

1. D. Helbing, S. Balietti, Fundamental and real-world challenges in economics. Sci. Cult. **76**(9–10), 399–417 (2010). http://www.saha.ac.in/cmp/camcs/Sci_Cul_091010/1720%20Dirk%20Helbing.pdf

2. P. Ormerod, D. Helbing, Back to the drawing board for macroeconomics. in *Whats the Use of Economics? Teaching the Dismal Science after the Crisis*, ed. D. Coyle (London Publishing Partnership, 2012), http://volterra.co.uk/wp-content/uploads/2013/03/2_Back-to-the-Drawing-Board-for-Macroeconomics.pdf

3. D. Helbing, A. Kirman, Rethinking economics using complexity theory. Real-World Econ. Rev. **64**, 23–52 (2013), http://www.paecon.net/PAEReview/issue64/HelbingKirman64.pdf; http://futurict.blogspot.ch/2013/04/how-and-why-our-conventional-economic_8.html

4. D. Helbing, Economics 2.0: The natural step towards a self-regulating, participatory market society. Evol. Inst. Econ. Rev. **10**, 3–41 (2013), https://www.jstage.jst.go.jp/article/eier/10/1/10_D2013002/_pdf; http://www.todayonline.com/singapore/new-kind-economy-born

5. D. Helbing, Globally networked risks and how to respond. Nature. **497**, 51–59 (2013), http://www.researchgate.net/publication/236602842_Globally_networked_risks_and_how_to_respond/file/60b7d52ada0b3d1494.pdf; http://www.sciencedaily.com/releases/2013/05/130501131943.htm

6. D. Helbing, Systemic Risks in Society and Economics. International Risk Governance Council (irgc). (2010), http://www.researchgate.net/publication/228666065_Systemic_risks_in_society_and_economics/file/9fcfd50bafbc5375d6.pdf

7. OECD Global Science Forum, Applications of Complexity Science for Public Policy: New Tools for Finding Unanticipated Consequences and Unrealized Opportunities (2008), http://www.oecd.org/science/sci-tech/43891980.pdf

8. The FuturIcT Knowledge Accelerator: Unleashing the Power of Information for a Sustainable Future, http://papers.ssrn.com/sol3/papers.cfm?abstract_id=1597095; http://arxiv.org/abs/1304.0788

9. D. Helbing, S. Balietti, et al., Visioneer special issue: How can we Learn to Understand, Create and Manage Complex Techno-Socio-Economic Systems? (2011), http://epjst.epj.org/index.php?option=com_toc&url=/articles/epjst/abs/2011/04/contents/contents.html

10. D. Helbing, S. Balietti, Big Data, Privacy, and Trusted Web: What Needs to Be Done, http://papers.ssrn.com/sol3/papers.cfm?abstract_id=2322082

11. J. van den Hoven, et al., FuturICT—The road towards ethical ICT. Eur. Phys. J. Spec. Top. **214**, 153–181 (2012), http://link.springer.com/article/10.1140/epjst/e2012-01691-2#page-1

12. S. Buckingham Shum, et al., Towards a global participatory platform. Eur. Phys. J. Spec. Top. **214**, 109-1-52 (2012), http://link.springer.com/article/10.1140/epjst/e2012-01690-3#page-1

13. For an overview of the Snowden revelations, http://www.theguardian.com/world/the-nsa-files

14. Heise Online, Bundesprsident Gauck "sehr beunruhigt" ber US-Überwachung (July 25, 2013), http://www.heise.de/newsticker/meldung/Bundespraesident-Gauck-sehr-beunruhigt-ueber-US-Ueberwachung-1924026.html; for further interesting quotes see http://www.spiegel.de/international/europe/eu-officials-furious-at-nsa-spying-in-brussels-and-germany-a-908614.html

15. D. Helbing, What the Digital Revolution Means for Us, Science Business (June 12, 2014), http://www.sciencebusiness.net/news/76591/What-the-digital-revolution-means-for-us, see also [16]

16. The Intercept, Barack Obamas secret terrorist-tracking system, by the numbers (August 5, 2014), https://firstlook.org/theintercept/article/2014/08/05/watch-commander/

17. Foreign Policy, 3 degrees of separation is enough to have you watched by the NSA (July 17, 2013), http://complex.foreignpolicy.com/posts/2013/07/17/3_degrees_of_separation_is_enough_to_have_you_watched_by_the_nsa; see also "Three degrees of separation" in http://www.theguardian.com/world/interactive/2013/nov/01/snowden-nsa-files-surveillance-revelat-ions-decoded#section/1

18. The Washington Post, NSA phone record collection does little to prevent terrorist attacks, group says (January 12, 2014), http://www.

washingtonpost.com/world/national-security/nsa-phone-record-collection-does-little-to-prevent-terrorist-attacks-group-says/2014/01/12/8aa860aa-77dd-11e3-8963-b4b654bc-c9b2_story.html?hpid=z4; http://securitydata.newamerica.net/nsa/analysis

19. M. Gill, Spriggs: Assessing the impact of CCTV. Home Office Research, Development and Statistics Directorate (2005), https://www.cctvusergroup.com/downloads/file/Martin%20gill.pdf; see also BBC News (August 24, 2009), 1000 cameras 'solve one crime', http://news.bbc.co.uk/2/hi/uk_news/england/london/8219022.stm

20. Home Office, Review of the Operation of Schedule 7 (September 2012), https://www.gov.uk/government/uploads/system/uploads/attachment_data/file/157896/consultation-do-cument.pdf; http://www.theguardian.com/commentisfree/2013/aug/18/david-miranda-detained-uk-nsa

21. National Geographic, The war on drugs is a "miserable failure" (January 22, 2013), http://newswatch.nationalgeographic.com/2013/01/22/the-war-on-drugs-is-a-miserable-failure/

22. Electronic Frontier Foundation, DEA and NSA team up to share intelligence, leading to secret use of surveillance in ordinary investigations (August 6, 2013), https://www.eff.org/deeplinks/2013/08/dea-and-nsa-team-intelligence-laundering

23. The Guardian, Eric Holder unveils new reforms aimed at curbing US prison population (August 12, 2013), http://www.theguardian.com/world/2013/aug/12/eric-holder-smart-crime-reform-us-prisons

24. The Intercept, Guilty until proven innocent (March 7, 2014), https://firstlook.org/theintercept/document/2014/03/07/guilty-proven-innocent/; http://www.huffingtonpost.com/2014/08/15/unlawful-arrests-police_n_5678829.html

25. J. Schmieder, Mit einem Bein im Knast (2013), http://www.amazon.com/Mit-einem-Bein-Knast-gesetzestreu-ebook/dp/B0-0BOAFXKM/ref=sr_1_1?ie=UTF8; http://www.spiegel.tv/filme/magazin-29122013-verboten/

26. Spiegel, Redtube.com: Massenabmahnungen wegen Porno-Stream (December 9, 2013), http://www.spiegel.de/netzwelt/web/porno-seite-redtube-abmahnungen-gegen-viele-nutzer-a-938077.html; http://www.

spiegel.de/netzwelt/web/massenabm-ahnungen-koennen-laut-gerichtsurteil-ein-rechtsmissbrauch-sein-a-939764.html

27. International Business Times, Microsoft tip leads to child porn arrest; Google, Facebook also scan for vile images (August 7, 2014), http://www.ibtimes.com/microsoft-tip-leads-child-porn-arrest-google-facebook-also-scan-vile-images-1651756

28. MailOnline, The terrifying rise of cyber crime: Your computer is currently being targeted by criminal gangs looking to harvest your personal details and steal your money (January 12, 2013), http://www.dailymail.co.uk/home/moslive/article-2260221/Cy-ber-crime-Your-currently-targeted-criminal-gangs-looking-steal-money.html; https://firstlook.org/theintercept/article/2014/03/12/nsa-plans-infect-millions-computers-malware/

29. New York Times, Russian Hackers Amass Over a Billion Internet Passwords (August 5, 2014), http://www.nytimes.com/2014/08/06/technology/russian-gang-said-to-amass-more-than-a-billion-stolen-internet-credentials.html?_r=0

30. Spiegel Online, Umfrage zum Datenschutz: Online misstrauen die Deutschen dem Staat (June 5, 2014), http://www.spiegel.de/netzwelt/web/umfrage-deutsche-misstrauen-dem-staat-beim-online-datenschutz-a-973522.html

31. Focus Money Online, Test enthüllt Fehler in jeder zweiten Schufa-Auskunft (August 8, 2014), http://www.focus.de/finanzen/banken/ratenkredit/falsche-daten-teure-gebuehren-te-st-enthuellt-fehler-in-jeder-zweiten-schufa-auskunft_id_40469-67.html

32. Versicherungsbote, Wirtschaftsspionage durch amerikanischen Geheimdienst NSA—Deutsche Unternehmen sind besorgt (July 2, 2013), http://www.versicherungsbote.de/id/89486/Wirtschaftsspionage-durch-amerikanischen-Gehe-imdienst-NSA/; http://pretioso-blog.com/der-fall-enercon-in-der-ard-wirtschaftsspionage-der-usa-durch-die-nsa-in-deutschla-nd-jedes-unternehmen-ist-betroffen/ and http://www.tagesschau.de/wirtschaft/wirtschafsspionage100.html

33. Zeit Online, Blackout (April 17, 2014), http://www.zeit.de/2014/16/blackout-energiehacker-stadtwerk-ettlingen

34. Stuxnet, http://en.wikipedia.org/wiki/Stuxnet; http://www.zdnet.com/blog/security/stuxnet-attackers-used-4-windows-ze-ro-day-exploits/7347; http://www.itworld.com/security/28155-3/researcher-warns-stuxnet-flame-show-microsoft-may-have-been-infiltrated-nsa-cia

35. PC News, Researchers warn about 'BadUSB exploit' (July 31, 2014), http://www.pcmag.com/article2/0,2817,2461717,00.asp

36. Mail Online, China is spying on you through your KETTLE: Bugs that scan wi-fi devices found in imported kitchen gadgets (October 31, 2013), http://www.dailymail.co.uk/news/article-2480900/China-spying-KETTLE-Bugs-scan-wi-fi-devices-imported-kitchen-gadgets.html

37. The Verge, Secret program gives NSA, FBI backdoor access to Apple, Google, Facebook, Microsoft data (June 6, 2013), http://www.theverge.com/2013/6/6/4403868/nsa-fbi-mine-data-apple-google-facebook-microsoft-others-prism; http://techrights.org/2013/06/15/nsa-and-microsoft/, http://techrights.org/2013/08/22/nsa-back-doors-blowback/

38. MIT Technology Review, NSAs own hardware backdoors may still be a "Problem from hell" (October 8, 2013), http://www.technologyreview.com/news/519661/nsas-own-hardware-backdoors-may-still-be-a-problem-from-hell/; http://www.theguardian.com/world/2013/sep/05/nsa-gchq-encryption-codes-security, http://www.eteknix.com/expert-says-nsa-have-backdoors-built-into-intel-and-amd-processors/, http://en.wikipedia.org/wiki/NSA_ANT_catalog

39. RT, NSA sued for hoarding details on use of zero day exploits (July 3, 2014), http://rt.com/usa/170264-eff-nsa-lawsuit-0day/; http://www.wired.com/2014/04/obama-zero-day/

40. Private Wifi, New drone can hack into your mobile device (March 31, 2014), http://www.privatewifi.com/new-drone-can-hack-into-your-mobile-device/; http://www.privatewifi.com/new-drone-can-hack-into-your-mobile-device/, http://securitywatch.pcmag.com/hacking/314370-black-hat-intercepting-calls-and-cloning-phones-with-femtocells, http://www.npr.org/blogs/alltechconsidered/2013/07/15/201490397/How-Hackers-Tapped-Into-My-Verizon-Cellphone-For-250, http://www.alarmspy.com/phone_hacking_9.html

41. The Guardian, XKeyscore: NSA tool collects 'nearly everything' a user does on the internet (July 31, 2013), http://www.theguardian.com/world/2013/jul/31/nsa-top-secret-program-online-data; http://en.wikipedia.org/wiki/XKeyscore

42. Business Insider, How a GED-holder managed to get 'top secret' government clearance (June 10, 2013), http://www.businessinsider.com/edward-snowden-top-secret-clearance-nsa-whistleblower-2013-6

43. The Guardian, Academics criticise NSA and GCHQ for weakening online encryption (September 16, 2013), http://www.theguardian.com/technology/2013/sep/16/nsa-gchq-undermine-internet-security

44. BBC News, Heartbleed bug: What you need to know (April 10, 2014), http://www.bbc.com/news/technology-26969629; see also http://en.wikipedia.org/wiki/Heartbleed

45. Huff Post, Apple May Be Spying On You Through Your iPhone (July 25, 2014), http://www.huffingtonpost.com/2014/07/26/apple-iphones-allow-extra_n_5622524.html; http://tech.firstpost.com/news-analysis/chinese-media-calls-apples-iphone-a-national-security-concern-227246.html

46. The Guardian, Chinese military officials charged with stealing US data as tensions escalate (May 20, 2014), http://www.theguardian.com/technology/2014/may/19/us-chinese-military-officials-cyber-espionage; http://www.nytimes.com/2006/06/07/washington/07identity.html, http://blog.techgenie.com/editors-pick/data-theft-incidents-to-prevent-or-to-cure.html, http://articles.economictimes.indiatimes.com/keyword/data-theft/recent/5

47. See http://en.wikipedia.org/wiki/Cyberwarfare and http://en.wikipedia.org/wiki/Cyber-attack and http://www.wired.com/2013/06/general-keith-alexander-cyberwar/all/

48. The National Interest, Cyberwar and the nuclear option (June 24, 2013), http://nationalinterest.org/commentary/cyberwar-the-nuclear-option-8638

49. Wired, Meet MonsterMind, the NSA Bot That Could Wage Cyberwar Autonomously (August 13, 2014), http://www.wired.com/2014/08/nsa-monstermind-cyberwarfare/

50. InfoWorld, The real story of how big data analytics helped Obama win (February 14, 2013), http://www.infoworld.com/d/big-data/the-real-story-of-how-big-data-analytics-helped-obama-win-212862; http://www.technologyreview.com/featuredstory/509026/how-obamas-team-used-big-data-to-rally-voters/; Nature, Facebook experiment boosts US voter turnout (September 12, 2012), http://www.nature.com/news/facebook-experiment-boosts-us-voter-turnout-1.11401; http://www.sbs.com.au/news/article/2012/09/13/us-election-can-twitter-and-facebook-influence-voters

51. RT, Western spy agencies build 'cyber magicians' to manipulate online discourse (February 25, 2014), http://rt.com/news/five-eyes-online-manipulation-deception-564/; https://firstlook.org/theintercept/2014/02/24/jtrig-manipulation/, https://firstlook.org/theintercept/2014/07/14/manipulating-online-polls-ways-british-spies-seek-control-internet/, http://praag.org/?p=13752

52. NZZ, Ins Amt gezwischert? (August 18, 2014) www.nzz.ch/aktuell/startseite/ins-amt-gezwitschert-1.18202760

53. Secret services are there to stabilize democracies? The reality looks different, https://www.facebook.com/FuturICT/posts/576176715754340

54. DW, Bulgarians protest government of 'oligarchs' (June 26, 2013), http://www.dw.de/bulgarians-protest-government-of-oligarchs/a-16909751; Tagesschau.de, Zorn vieler Bulgaren ebbt nicht ab (July 24, 2013), http://www.tagesschau.de/ausland/bulgarienkrise102.html

55. Mit dem Recht auf informationelle Selbstbestimmung wären eine Gesellschaftsordnung und eine diese ermöglichende Rechtsordnung nicht vereinbar, in der Bürger nicht mehr wissen können, wer was wann und bei welcher Gelegenheit über sie weiß. Wer unsicher ist, ob abweichende Verhaltensweisen jederzeit notiert und als Information dauerhaft gespeichert, verwendet oder weitergegeben werden, wird versuchen, nicht durch solche Verhaltensweisen aufzufallen. […] Dies würde nicht nur die individuellen Entfaltungschancen des Einzelnen beeinträchtigen, sondern auch das Gemeinwohl, weil Selbstbestimmung eine elementare Funktionsbedingung eines auf Handlungsfähigkeit und Mitwirkungsfähigkeit seiner Bürger begründeten freiheitlichen demokratischen Gemeinwesens ist. Hieraus folgt: Freie Entfaltung

der Persönlichkeit setzt unter den modernen Bedingungen der Daten-verarbeitung den Schutz des Einzelnen gegen unbegrenzte Erhebung, Speicherung, Verwendung und Weitergabe seiner persönlichen Daten voraus. Dieser Schutz ist daher von dem Grundrecht des Art. 2 Abs. 1 in Verbindung mit Art. 1 Abs. 1 GG umfasst. Das Grundrecht gewährleistet insoweit die Befugnis des Einzelnen, grundsätzlich selbst über die Preisgabe und Verwendung seiner persönlichen Daten zu bestimmen. See also Alexander Rossnagel (August 28, 2013) "Big Data und das Konzept der Datenschutzgesetze", http://www.privacy-security.ch/2013/Download/Default.htm

56. Foreign Policy, The Social Laborary (2014), http://www.foreignpolicy.com/articles/2014/07/29/the_social_laboratory_singapore_surveillance_state; also see D. Helbing, Google as God? Opportunities and Risks of the Information Age (March 27, 2013), http://futurict.blogspot.ie/2013/03/google-as-god-opportunities-and-risks.html; see also From crystal ball to magic wand: The new world order in times of digital revolution, https://www.youtube.com/watch?v=AErRh_yDr-Q

57. See the videos, http://www.youtube.com/watch?v=I_Lphxknozc and http://www.youtube.com/watch?v=AErRh_yDr-Q

58. live science, U.S. torture techniques unethical, ineffective (January 6, 2011), http://www.livescience.com/9209-study-torture-techniques-unethical-ineffective.html; http://en.wikipedia.org/wiki/Effectiveness_of_torture_for_interrogation and http://www.huffingtonpost.com/2014/04/11/cia-harsh-interrogations_n_5130218.html; The Guardian, US drone attacks 'counter-productive', former Obama security adviser claims (January 7, 2013), http://www.theguardian.com/world/2013/jan/07/obama-adviser-criticises-drone-policy; http://www.huffingtonpost.com/2013/05/21/us-drone-strikes-ineffective_n_3313407.html and http://sustainablesecurity.org/2013/10/24/us-drone-strikes-in-pakistan/; NationalJournal, The NSA isn't just spying on us, it's also undermining Internet security (April 30, 2014), http://www.nationaljournal.com/daily/the-nsa-isn-t-just-spying-on-us-it-s-also-undermining-internet-security-20140429; http://www.slate.com/blogs/future_tense/2014/07/31/usa_freedom_act_update_how_the_nsa_hurts_our_economy_cybersecurity_and_foreign.html

59. D. Helbing, et al., Saving human lives: What complexity science and information systems can contribute. J. Stat. Phys. (2014), http://link. springer.com/article/10.1007%2Fs10955-014-1024-9

60. Time, S&P: Income Inequality Is Damaging the Economy (August 5, 2014), http://time.com/3083100/income-inequality/

61. physicstoday, Qualified trust, not surveillance, is the basis of a stable society (July 2013), http://scitation.aip.org/content/aip/magazine/ physicstoday/news/10.1063/PT.4.2508; the Foreword in Consumer Data Privacy in a Networked World, (February 2012) http://www. whitehouse.gov/sites/default/files/privacy-final.pdf, which starts: "Trust is essential to maintaining the social and economic benefits that networked technologies bring to the United States and the rest of the world."

62. World Economic Forum, Personal Data: Emergence of a New Asset Class, (2011) http://www.weforum.org/reports/personal-data-emergence-new-asset-class

63. Y.-A. de Montjoye, E. Shmueli, S.S. Wang, A.S. Pentland, openPDS: Protecting the Privacy of Metadata through SafeAnswers (2014), http:// www.plosone.org/article/info%3Adoi%2F10.1371%2Fjournal.pone. 0098790; ftp://131.107.65.22/pub/debull/A12dec/large-scale.pdf, http://infoclose.com/protecting-privacy-online-new-system-would-give-individuals-more-control-over-shared-digital-data/, http://www.taz. de/!131892/, http://www.taz.de/!143055/

64. D. Helbing, The world after Big Data: Building the self-regulating society (August 14, 2014), https://www.youtube.com/watch?v=I_Q_-Pk-btY or https://www.youtube.com/watch?v=I_Lphxknozc

2
Lost Robustness

This chapter first appeared in the NAISSANCE Newsletter under the title "Lost Robustness" and is reproduced here with kind permission of Naissance Capital and with minor stylistic improvements.

The current financial crisis is the expression of a systemic change that has occurred in the global economy slowly but profoundly during the last few decades. Our thesis results from an analysis of the financial world from the perspective of the theory of complex systems (which describes common features of social, traffic, ecnomic and ecological systems). The key question guiding our analysis is: what properties make the financial system robust, and therefore stable?

Is the present financial crisis different from preceding ones? Many different answers are given to this question. However, what is more important than the answers themselves is the validity of the theoretical concepts, on which they are based. It is obvious that the prevalent mathematical models of economic processes and the risks attributed to them have failed—a failure that happened just at the time when they would have been most needed. In our opinion, it is not only the faulty use of risk evaluation models that underlies this failure, but a wrong description of the financial system itself. Since the early 1980s, the financial system has fundamentally changed, which must be viewed as a systemic change of financial markets. Keywords here are the creation of new financial instruments (e.g. derivatives), an acceleration on all process levels, and the removal of obstacles that formerly impeded international capital flows.

Although the present analysis of the crisis touches these issues, it does not well enough elaborate its systemic aspects, leaving it inadequate and incomplete. In our opinion, it is necessary to examine the robustness of the financial system. Robustness is a concept from complexity theory, which in general focuses on universal patterns of behavior and development in complex systems. We want to show that this approach is useful for an understanding of the current crisis, because it illustrates that recent changes in the financial system have seriously undermined its robustness.

2.1   Understanding Complex Systems

Complex systems are characterized by numerous interacting actors and factors. Examples are social, economic, or traffic systems, as well as the behavior of crowds or ecosystems. The behavior of these systems is often dominated by their internal dynamics. Attempts to control them from outside frequently lead to unexpected and unintended results. Insights into the behavior of such systems, however, can be gained by a three-step procedure. First, the actors of the system and their interactions must be captured as completely as possible and mapped into a network of causal interdependencies. Second, one needs to determine the properties of the system that are most relevant for its state and dynamics. By specifying favourable states of the system, one can introduce a normative element into the model. Third, one must identify the most important external influences, which affect the system, but are not influenced by it.

Many complex systems analyzed in this way—both natural and artificial ones—show common general properties: For example, there are phases during which the overall behavior of the system remains more or less unchanged, although interactions within the system as well as external influences may fluctuate considerably. When this is the case, the system is stable—not in the sense that nothing happens, but that changes are restricted and foreseeable, even if this is not intended by the individual actors. (This system behavior may be compared with Adam Smith's principle of the invisible hand.) However, if certain external or internal conditions of the system change beyond a certain threshold, the stability can collapse. In this case, the system behavior becomes massively different, and one speaks of a regime shift. For example, established actors may disappear or new ones may emergence. The corresponding systemic processes often occur in a cascade- or avalanche-like manner, and the frequency of such extreme events is much higher than expected. It typically follows a so-called fat-tailed distribution rather than a normal distribution, while the normal distribution is still the basis of most risk assessment models today.

The ability of a system to avoid such regime shifts is called robustness. The issue of robustness is central to an analysis of the financial system, as it is often the interactions in a system which determine its robustness. Moreover, in contrast to external factors, interactions are susceptible to regulation and, thus, to human intervention. In order to understand robustness, it is helpful to consider the stability properties of other complex systems, for example, ecological systems. During evolution, these systems have been exposed to numerous disturbances and nevertheless (or, in some way, exactly for this reason) achieved an amazing stability. Five key properties of complex systems have proved to be advantageous for robustness: variety, redundancy, compartmentalization, sparseness (i.e. a low degree of interconnectedness), and mutually adjusted time scales of processes in the system.

Variety, i.e. the existence of different kinds of actors and the application of different strategies in the system, enhances the adaptability and guarantees that not all actors face a stability crisis at the same time. Redundancy allows the system to deal with the loss of system components (e.g. actors, resources, or process pathways). To put it simple, several safeguards must fail, before a problem develops into a crisis. Should the trouble nevertheless skyrocket, e.g. due to an unfortunate coincidence of several problems, compartmentalization can help: If the network underlying the system (such as the network of interbank loans) is subdivided into barely interdependent subnetworks, this supports a decoupling of the system into autonomous subsystems when needed. In some sense, compartementalization introduces predetermined breaking points, i.e. the idea is to transfer the principle of fuses in electrical networks to prevent serious damage in other networks as well. Such a strategy could avoid an epidemic spreading of trouble across the entire system. Demanding sparseness complies with the principle of decoupling, but it also considers the fact that interactions between actors are costly. A system with an excessive number of connections is, therefore, inefficient. Finally, adjusted time scales guarantee that the processes in the system are well coordinated, so that they do not disturbe each other.

2.2   Criticality and Lack of Transparency

In complex systems, several destabilizing factors may be at work. Of particular importance in this connection is an effect called self-organized criticality. Other relevant factors are time delays and the lack of transparency as a result of growing system complexity. Moreover, in the case of the financial system, market mechanisms themselves imply a certain degree of instability, as we will show.

Self-organized criticality means that the system maneuvers itself into a critical state. In case of the financial system, this results from the fact that banks must compete for customers, which forces them to take greater and greater risks. Hence, unexpected economic developments could eventually get some banks into trouble, and the corresponding enterprises would go bankrupt or would be taken over, which is a natural adjustment process in a free market economy. However, the process may be dangerous when many actors simultaneously move towards a critical state, which has indeed been the case in the current crisis. For example, banks invested into mortgage-backed securities that promised higher net yields, but the more mortgages were taken out, the more house prices were driven up. Prices in the American housing market roughly doubled from 2000 to 2006, while US wages increased only around 14 % in the same period. It is clear that many people could not pay for housing at some point and that the real-estate bubble would burst—a typical example of self-organized criticality.

The dangers of self-organized criticality can be counteracted by the above-mentioned strategy of compartmentalization. Fighting forest fires is a good analogy. On one hand, in areas of low intervention, forest fires are frequent but spatially limited, as previous fires create natural firebreaks. On the other hand, suppressing fires in an early stage prevents many fires from spreading. However, once they manage to spread, they often become uncontrollable and result in a large-scale conflagration, because there are no natural firebreaks. The Feds monetary policy in the last few years could be interpreted in that way: It prevented the formation of natural firebreaks in the financial system. Furthermore, allowing the packaging of mortgages and their global disposal permitted around 40 % of the values related to these mortgages to circumvent government regulations. By this, the occurrence of fires in the financial market was not even monitored, i.e. there was a serious lack of transparency.

In this context, one should be aware that, in socio-economic systems, the lack of transparency is growing naturally. The evolution of complex systems is characterized by an increasing degree of differentiation, which facilitates new structures and processes— and consequently requires new experts that are able to overlook the details of a more and more complex world. This process can be readily observed in the globalization of the economy, in legislation and jurisdiction, and in the financial market. The increasing level of complexity, however, eventually reduces the transparency of the system and, thereby, its controllability. An example for this is the creation of new and ever more complex financial products (derivatives) that were built on previously existing ones. This indeed opened up new opportunities, as reflected by the enormous expansion of the values represented in derivatives: In the spring of 2008, derivative contracts amounted to more than $ 500 trillion (two orders of magnitude more than in 1980). However, among these were also credit-default swaps, which did not exist a few years ago and now amounted to $ 45–60 trillion (depending on source), while they were barely attributable to real-world economic goods—an obvious problem of transparency. Moreover, the fact that financial instruments became more and more complex became particularly problematic as, with regard to expertise, the robustness criteria of variety (in terms of different approaches to risk assessment) and redundancy (requiring that different experts investigate the same risks) were compromised: In the last three decades, basically three enterprises—Standard & Poor's, Moody's and Fitch Ratings—with a worldwide market-share of more than 90 % have rated financial assets and valuated the underlying real-world economic goods. This, of course, created massive correlations and serious herding effects among financial actors.

2.3   Acceleration and De-Compartmentalization

If the time scales of different processes in complex systems do not match well, a variety of delay-related problems may occur, which is reflected in financial systems on different levels. Let us first focus on short time scales: When the stock market crashed, computers sometimes did not catch up with all the requested transactions in real time anymore. As a consequence, some orders were executed with delays, which affected the efficiency of real-time trading. However, when the equilibration to the fundamental value is delayed, this may induce over-reaction of market participants and overshoots in the market, i.e. temporary disequilibria. In traffic flows, such delays can lead to so-called phantom traffic jams, i.e. breakdowns of free flow for no obvious reasons (such as accidents or bottlenecks would be). Why should such unexpected breakdown effects not show up in financial markets as well?

On a longer time-scale one finds that it took months to gain at least a partial overview of the risks and losses in the financial system. This reflects the large degree of interconnectedness, i.e. de-compartetmentalization, which has occurred. Since the beginning of this decade, new financial products were developed at a rapid pace. One example are securitization products, which are composed by gathering a group of debt obligations, such as mortgages, into a pool, and then dividing that pool into portions that can be sold as securities in the secondary market. In this way, particularly due to the use of multi-level securitizations, it became decreasingly transparent, which real-world goods were actually behind these securities. In the end, one had to rely on the affirmations of rating agencies. Their operation, however, was based on the intuition that the risk could be dispersed by distributing it over different areas of economy. That approach, however, has rather created conditions which allowed for an unobstructed spreading of the crisis.

This is an important point. All previous economic crises were clearly limited in terms of the geographical region (e.g. the Asia crisis affecting the rising tiger states in 1997/1998) or the economic sector concerned (e.g. in case of the Dotcom-bubble in 2000). Within such compartments, euphoria and panic could occur time and again—but they were contained. In our current financial system, however, changes in state regulation undermined compartmentalization. An example is the American Glass-Steagall Act which, until it was abolished in 1999, had prevented companies from doing both commercial and investment banking. Compartmentalizations like these are not only objectively important, but also stabilizing from a psychological point of view: In times of crisis, human decision-making tends to be rapid and based on intuitions oriented at clearly perceivable structures of economic reality. Compartmentalization is also a precondition to allow for effective state legislation and to ensure, in times of crisis, that the process of re-evaluation does not take too long.

2.4   Systemic Stability and Trust

Apart from the destabilizing factors discussed so far, financial markets have an additional, built-in destabilizing mechanism. The possibility of profiting not only from rising stock prices but also from falling ones (by short-selling) has important consequences. In principle, participants can earn faster and more money when the market goes up and down, as compared to a stable, steady-growing market. Hence, opportunities for profits grow with market variability, which is likely to increase the volatility of stock prices. Therefore, market mechanisms are not constructed in a way that would support stable stock prices. They rather imply an inherent instability that needs to be continuously counterbalanced. This may be compared with human walking, which can be characterized by continuously counteracting the process of falling. If this control process of continuous counter-action fails, a free fall results, as has been recently observed in the financial system.

Such systemic instability is, of course, potentially more harmful, the smaller the financial safety margins of the involved actors are. In fact, compared to the 1980s, there has been a significant change in this regard, for example, in the United States: U.S. household debts, when measured in percentage of income, have quintupled to reach a level of 130 %. The debt of US banks increased up to 110 % of GDP. Given these conditions, changes in the value of enterprises as a consequence of the systemic instability of financial markets can have serious effects.

What makes things worse is the fact that the auction mechanism used in financial markets mixes material values with psychology. As we have seen, if actors lose interest in financial transactions, prices can fall indefinitely. It is, therefore, important to consider the social network underlying the valuation of real-world goods: Prices in a market economy also reflect the consumers trust (or distrust) with regard to the usefulness of the related good. This in-priced trust is also the reason why, mathematically speaking, economic transactions are not a zero-sum game, i.e. there are no conservation laws for economic values comparable to the law of energy conservation in physics. The valuation process of economic goods is linked to a social network of interacting actors, where the depth and duration of the interaction correlates with the degree of trust between them. We may call this a network of trust. The successful build-up of a trust network requires a minimum degree of intimacy between the actors. Trust links a real-world economic good with its price, which in this way becomes a tradable object in the financial system.

However, also the trust network has been negatively affected in the past years. For numerous reasons, the frequency and duration of interactions during the process of valuating economic goods were reduced: Geographical distances between business partners increased, time constraints were imposed on business transactions, second opinions were not anymore obtained for financial reasons, language barriers became an issue in a globalized economy, incentive structures within companies provoked more rapid staff turnover, etc. The fact that investors currently hold stocks of Nasdaq-quoted enterprises only for two months on average, while in the 1980s, this time period was about four years - is only one indicator of this problem: Many companies are confronted with a new ownership structure at the time of each quarterly report even in research-intensive sectors, where companies are required to make long-term investments.

2.5 Utilizing Control Features of Complex Systems

An analysis of the financial system from the viewpoint of the theory of complex systems, therefore, reveals a much more differentiated picture than most public debates of the crisis do. The behavior of financial markets in early 2008 shows all the signs of a conflagration that has gone out of control. The continuing erosion of banks' reserves shows that it has become unclear what a subprime mortgage actually is and, therefore, even the name of this crisis is quite misleading. Therefore, it is not surprising that in early 2008, many different estimates regarding the overall need for depreciation have been circulated—from $ 170 billion (Bank of England) over $ 400 billion (OECD) up to $ 1 trillion (IMF) or even more ($ 4 trillion, Goldman Sachs). These estimates differ by more than one order of magnitude (compare also to the actual estimates in the previous article).

We therefore conclude that discussing whether capital injections by the central banks will be useful or not, and how large they should be, is by far too short-sighted. Capital injections fertilize a system that already has lost its stability to a large extent. The question of how to control the system must rather focus on the properties that are relevant for its robustness. Our analysis of the crucial factors for robustness offers a new approach to a better understanding of complex systems like the financial one, and to a mitigation of crises or, ideally, even an avoidance of them. This, however, requires a considerable expansion of currently available research capacities. ETH Zurich has, therefore, recently set up a competence center for Coping with Crises in Complex Socio-Economic Systems. It unites researchers from the economic, social, natural and engineering sciences, who are committed to the modeling of social and economic systems, and to the development of methods for their stabilization.

2.6   Author Information

R. James Breiding is co-founder and manager of the Zurich-based investment company Naissance Capital Ltd.; Markus Christen works in the research team of Naissance Capital and investigates empirical aspects of moral behavior at the University of Zurich. Dirk Helbing is professor of sociology, in particular of modeling and simulation at ETH Zurich - Swiss Federal Institute of Technology, and leads the competence center Coping with Crises in Complex Socio-Economic Systems. This text is a translation of: R. J. Breiding, M. Christen, D. Helbing: Ist das globale Finanzsystem noch robust?, written in March 2008.

3

How and Why Our Conventional Economic Thinking Causes Global Crises

This chapter first appeared as FuturICT blog on April 8, 2013, see http://futurict.blogspot.de/2013/04/how-and-why-our-conventional-economic_8.html, and is reproduced here with minor stylistic improvements. An extended version has been published as a paper by Dirk Helbing and Alan Kirman (2013) Rethinking economics using complexity theory. Real-World Economics Review 64, see http://www.paecon.net/PAEReview/issue64/HelbingKirman64.pdf.

This discussion paper challenges a number of established views of mainstream economic thinking that, from the perspective of complexity science, seem to require a thorough revision. As Albert Einstein pointed out: "We cannot solve our problems with the same kind of thinking that created them." Therefore, the new perspective offered here might help to identify new solutions to a number of old economic problems.

I believe it's no wonder that our world is in trouble. We currently lack the global systems science that would allow us to understand the world, which is now changing more rapidly than we can collect the experience required to cope with upcoming problems. We also cannot trust our intuitions, since the complex systems we have created behave often in surprising, counter-intuitive ways. Frequently, their properties are not determined by their components, but by their interactions. Therefore, a strongly coupled world behaves fundamentally different from a weakly coupled world with independent decision-makers. Strong interactions tend to make the system uncontrollable—they create cascading effects and extreme events.

As a consequence of the transition to a more and more strongly coupled world, we need to revisit the underlying assumptions of the currently prevailing economic thinking. In the following, I will discuss ten widespread assertions, which would work in a perfect economic world with representative agents and uncorrelated decisions, where heterogeneity, decision errors, and time scales do not matter. However, they are apparently not well enough suited to depict the strongly interdependent, diverse, and quickly changing world, we are facing, and this has important implications. Therefore, we need to think outside the box and require a paradigm shift towards a new economic thinking characterized by a systemic, interaction-oriented perspective inspired by knowledge about complex, ecological, and social systems. As Albert Einstein noted, long-standing problems are rarely solved within the dominating paradigm. However, a new perspective on old problems may enable new mitigation strategies.

3.1 "More Networking Is Good and Reduces Risks"

Many human-made systems and services are based on networking. While some degree of networking is apparently good, too much connectivity may also create systemic risks and pathways for cascading effects. These may cause extreme events and global crises like the current financial crisis. Moreover, in social dilemma situations (where unfair behavior or cheating creates individual benefits), too much networking creates a breakdown of cooperation and trust, while local or regional interactions may promote cooperation. The transformation of the financial system into a global village, where any agent can interact with any other agent, may actually have been the root cause of our current financial crisis.

Countermeasures Limit the degree of networking to a healthy amount (e.g. by a link-based progressive tax) and/or introduce adaptive decoupling strategies to stop cascading effects and enable graceful degradation (including slow-down mechanisms in crisis situations). Support the evolution and co-existence of several weakly coupled financial systems (to reduce systemic vulnerability, stimulate competition between systems, and create backup solutions). Reduce the complexity of financial products and improve the transparency of financial interdependencies and over-the-counter transactions by creating suitable information platforms.

3.2 "The Economy Tends Towards an Equilibrium State"

Current economic thinking is based on the assumption that the economic system is in equilibrium or at least tends to develop towards a state of equilibrium. However, today's world changes faster than many companies and policies can adapt. Therefore, the world's economic system is unlikely to be in equilibrium at any point in time. It is rather expected to show a complex non-equilibrium dynamics.

Therefore, a new economic thinking inspired by complex dynamical systems, ecosystems, and social systems would be beneficial. Such a perspective would also have implications for the robustness of economic systems. Overall, beneficial properties seem to be: redundancy, variety, sparseness, decoupling (separated communities, niches), and mutually adjusted time scales (which are required for hierarchical structures to function well).

Countermeasures   Invest into new economic systems thinking. Combine the axiomatic, mathematical approach of economics with a natural science approach based on data and experiments. Develop non-equilibrium network models capturing the self-organized dynamics of real economic systems. Pursue an interdisciplinary approach, taking into account complex, ecological and social systems thinking. Develop better concepts for systemic risk assessment, systems design, and integrated risk management.

3.3   "Individuals and Companies Decide Rationally"

The homo economicus is a widely used paradigm in economics. It is the basis of a large and beautiful body of mathematical proofs on idealized economic systems. However, the paradigm of a strictly optimizing, perfect egoist is a model, which is questioned by theoretical and empirical results.

Theoretically, the paradigm assumes unrealistic information storage and processing capacities (everyone would need to have a full 1:1 representation of the entire world in the own brain and an instant data processing of huge amounts of data, including the anticipation of future decisions of others). Moreover, it has recently been found that not just a self-regarding homo economicus, but also an other-regarding homos socialis may result from the merciless forces of evolution. In fact, empirically one finds that people behave in a more cooperative and fair way than the paradigm of the homo economicus predicts. In particular, the paradigm neglects the role of errors, emotions, other-regarding preferences, etc. This implies significant deviations of real human behaviors from theoretically predicted ones.

Countermeasure Use a combination of interactive behavioral experiments, agent-based modeling, data mining, social supercomputing and serious multi-player on-line games to study (aspects of) real(istic) economic systems.

3.4 "Selfish Behavior Optimizes the Systemic Performance and Benefits Everyone"

Another pillar of conventional economic thinking is the principle of the invisible hand, according to which selfish profit maximization would automatically lead to the best systemic outcome based on self-organization. It is the basis of the ideology of homogeneous unregulated markets, according to which any regulation would tend to reduce the performance of economic systems.

However, models in evolutionary game theory show that self-organized coordination in markets can easily fail, even when market participants have equal power, symmetrical information etc. Moreover, even if the individually optimal behavior also maximizes system performance and if everybody behaves very close to optimal, this may still create a systemic failure (e.g. when the system optimum is unstable). Therefore, it is highly questionable whether the systemic inefficiencies resulting from competitive or uncoordinated individual optimization efforts can always be compensated for by greedy motivations (such as trying to get more than before or more than others).

Countermeasures Measure the system state in real-time and respond to this information adaptively in a way that promotes coordination and cooperation with the interaction partners. Create a 'Planetary Nervous System', i.e. an information and communication system supporting collective (self-)awareness of the impact of human actions on our world. Pluralistic reputation systems should be part of this. Increase opportunities for social, economic and political participation.

3.5 "Financial Markets Are Efficient"

One implication of the principle of the invisible hand is the efficiency of financial markets, according to which any opportunity to make money with a probability higher than chance would immediately be used, thereby eliminating such opportunities and any related market inefficiencies.

Efficient markets should not create bubbles and crashes, and therefore one would not need contingency plans for financial crises (they could simply not occur). Financial markets would rather be in equilibrium as the conventional Dynamic Stochastic General Equilibrium Models suggest. However, many people believe that bubbles and crashes do occur. Flash crashes are good examples for market inefficiencies, which have repeatedly occurred in the recent past. Also, many financial traders do not seem to believe in efficient markets, but rather in the existence of opportunities that can be used to make disproportional profits.

Countermeasures   Develop contingency plans for financial crises. Modify the financial architecture and identify suitable strategies (such as breaking points) to stop cascading effects in the financial system. Introduce noise into financial markets by random trading transactions to destroy bubbles before they reach a critical size that may have a disastrous systemic impact.

3.6 "More Information and Financial Innovations Are Good"

One common view is that market inefficiencies result from an unequal distribution of power, which partially results from information asymmetries (knowledge is power). Therefore, providing more information to everyone should remove the related inefficiencies.

However, too much information creates a cognitive information overload. As a result, people tend to orient at other people's behaviors and information sources they trust. As a consequence, people do no longer take independent decisions, which can undermine the "wisdom of crowds" and market efficiency. One example is the large and unhealthy impact that the assessments of a few rating agencies have on the global markets.

It is also believed that financial innovations will make markets more efficient by making markets more complete. However, it has been shown that complete markets are unstable. In fact, leverage effects, naked short-selling (of assets one does not own), credit default swaps, high-frequency trading and other financial instruments may have a destabilizing effect on financial markets.

Countermeasures Identify and pursue decentralized, pluralistic, participatory information platforms, which support the "wisdom of crowds" effect. Test financial instruments (such as derivatives) for systemic impacts (e.g. by suitable experiments and computer simulations) and certify them before they are released. Such certification is common in other economic sectors. (Special safety regulations apply, for example, in the electrical, automobile, pharmacy and food sectors.)

3.7 "More Liquidity Is Better"

Another wide-spread measure to cure economic crises are cheap loans provided by central banks. While this is intended to keep the economy running and to promote investments in the real economy, most of this money seems to go into financial speculation, since business and investment banks are not sufficiently separated.

This can cause bubbles in the financial and real-estate markets, where much of these cheap loans are invested. However, the high returns in the resulting bull markets are not sustainable, since they depend on the continued availability of cheap loans. Sooner or later, the created bubbles will implode and the financial market will crash (the likelihood of which goes up when the interest rates are increased). This again forces central banks to reduce interest rates to a minimum in order to keep the economy going and promote investments and growth. In other words, too much liquidity is as much of a problem, as is too little.

Countermeasure   Separate investment from business banks and introduce suitable adaptive transaction fees at financial markets.

3.8 "All Agents can Be Treated as if Acting the Same Way"

The 'representative agent approach' is another important concept of conventional economic thinking. Assuming that everyone would behave optimally, as the paradigm of the homo economicus predicts, in equivalent situations everybody should behave the same. It is therefore common to replace the interaction of an economic agent with other agents by interactions with average agents, in particularly if one assumes that everyone has access to the same information and participates in perfect markets.

However, the representative agent model cannot describe cascade effects well. These are not determined by the average stability, but by the weakest link. The representative agent approach also neglects effects of spatial interactions and heterogeneities in the preferences of market participants. When these are considered, the conclusions can be completely different, sometimes even opposite (e.g. there may be an outbreak rather than a breakdown of cooperative behavior).

Finally, the representative agent approach does not allow one to understand particular effects of the interaction network, which may promote or obstruct cooperativeness, trust, public safety, etc. Neglecting such network effects can lead to a serious underestimation of the importance of social capital for the creation of economic value and social well-being.

Countermeasures Protect economic and social diversity. Allow for the existence of niche markets and for the consideration of justified local advantages. Avoid competition on one single dimension (e.g. economic value generation) and promote multi-criterion incentive systems. Develop better compasses for decision making than GDP per capita, taking into account environmental, health, and social factors. Make social capital (such as cooperativeness, trust, public safety, ...) measurable.

3.9 "Regulation can Fix the Imperfections of Economic Systems"

When the self-organization of markets does not work perfectly, one often tries to fix the problem by regulation. However, complex systems cannot be steered like a bus, and many control attempts fail. In many cases, the information required to regulate a complex system is not available, and even if one had a surveillance system that monitors all variables of the system, one would frequently not know what the relevant control parameters are. Besides, suitable regulatory instruments are often lacking.

A more promising way to manage complexity is to facilitate or guide favorable self-organization. This is often possible by modifying the interactions between the system components. It basically requires one to establish targeted real-time information feedbacks, suitable rules of the game, and sanctioning mechanisms. To stay consistent with the approach of self-organization, sanctioning should as far as possible be done in a decentralized, self-regulatory way (as it is characteristic for social norms or the immune systems).

Countermeasures   Pursue a synergetic approach, promoting favorable self-organization by small changes in the interactions between the system elements, i.e. by fixing suitable rules of the game to avoid instabilities and suboptimal systemic states. (Symmetry, fairness, and balance may be such principles.) Introduce a global but decentralized and manipulation-resistant multi-criterion rating system, community-specific reputation system, and pluralistic recommender system encouraging rule-compatible behavior.

3.10 "Moral Behavior Is Good for Others, but Bad for Oneself"

Species that do not strictly optimize their benefits are often assumed to disappear eventually due to the principles of natural selection implied by the theory of evolution. As a consequence, a homo economicus should remain, while moral decision-making, which constrains oneself to a subset of available options, should vanish.

This problem certainly occurs, if one forces everybody to interact with everybody else on equal footing, as the concept of homogeneous markets demands. In fact, evolutionary game-theoretical models show that these are conditions under which a tragedy of the commons tends to occur, and where cooperation, fairness and trust tend to erode. On the other hand, social systems have found mechanisms to avoid the erosion of social capital. These mechanisms include repeated interactions, reputation effects, community interactions, group competition, sanctioning of improper behavior etc. In particular, decentralized market interactions seem to support fairness. Recent scientific breakthroughs even show that biological evolution can create a homo socialis' with other-regarding preferences.

Countermeasures Promote value-sensitive designs of monetary systems and of information and communication systems. Reputation systems, for example, would be an important element of these. They can also be used to define a new kind of money, so-called 'qualified money'. Moreover, it would be wise to introduce several co-existing, interacting, competitive exchange systems: one for anonymous (trans)actions (as we largely have them today) and one for accountable, traceable (trans)actions (creating social money or information). Additionally, one should create incentives for accountable, responsible (trans)actions and for ethical behavior.

3.11 Summary

We are now living in a strongly coupled and strongly interdependent world, which poses new challenges. While it is probably unrealistic to go back beyond the level of networking and globalization we have reached, there is a great potential to develop new management approaches for our complex world based on suitable interaction rules and adaptive concepts, using real-time measurements.

Our current financial and economic problems cannot be solved within the current economic mainstream paradigm(s). We need to change our perspective on the financial and economic system and pursue new policies. The following recommendations are made:

1. Adjust the perspective of our world to the fundamentally changed properties of the globalized, strongly interdependent techno-socio-economic-environmental system we have created and its resulting complex, emergent dynamic system behavior.
2. Make large-scale investments into new economic thinking, particularly multi-disciplinary research involving knowledge from sociology, ecology, and complexity science.
3. Support diversity in the system, responsible innovation, and multidimensional competition.
4. Recognize the benefits of local and regional interactions for the creation of social capital such as cooperativeness, fairness, trust, etc.
5. Require an advance testing of financial instruments and innovations for systemic impacts and restrict destabilizing instruments.
6. Identify and establish a suitable institutional framework for interactions (suitable rules of the game) in order to promote a favorable self-organization.
7. Implement better, value-sensitive incentive systems to foster more responsible action.
8. Establish a universal, global reputation system to promote fair behavior and allow ethical behavior to survive in a competitive world.

9. Create new compasses for political decision-making, considering environment, health, social capital, and social well-being.

10. Develop new tools to facilitate the assessment of likely consequences of our decisions and actions (the social footprint).

These tools may, for example, include

- a Planetary Nervous System to enable collective awareness of the state of our world and society in real-time,
- a Living Earth Simulator to explore side effects and opportunities of human decisions and actions,
- a Global Participatory Platform to create opportunities for social, economic and political participation,
- exchange systems that support value-oriented interactions.

The socio-economic system envisaged here is characterized by the following features: it is

1. based on individual decisions and self-organization,
2. using suitable incentives to support sustainability and to avoid coordination failures, tragedies of the commons, or systemic instabilities,
3. recognizing heterogeneity and diversity as factors promoting happiness, innovation, and systemic resilience.

Further Reading

1. D. Helbing, A. Kirman, Rethinking economics using complexity theory. Real-World Econ. Rev. **64**, 23–52 (2013)
2. D. Helbing, S. Balietti, Fundamental and real-world challenges in economics. Sci. Cult. **76**(9–10), 399–417 (2010)

3. D. Helbing, Accelerating scientific discovery by formulating grand scientific challenges. EPJ Special Top. **214**, 41–48 (2012)
4. T.C. Grund, C. Waloszek, D. Helbing, How natural selection can create both self-and other-regarding preferences, and networked minds. Sci. Reports **3**, 1480 (2013)
5. D. Helbing, Economics 2.0: the natural step towards a self-regulating, participatory market society. Evol. Institutional Econ. Rev. **10**(1), 1–39, see http://papers.ssrn.com/sol3/papers.cfm?abstract_id=2267697 (2013)
6. D. Helbing, A new kind of economy is born—social decision-makers beat the 'homo economicus', see http://papers.ssrn.com/sol3/papers.cfm?abstract_id=2332692 (2013)
7. D. Helbing, Globally networked risks and how to respond. Nature **497**, 51–59 (2013)

4

"Networked Minds" Require a Fundamentally New Kind of Economics

This chapter was first published on March 20, 2013, at http://www.alphagalileo.org/ViewItem.aspx?ItemId=129550& CultureCode=en and is reproduced here with minor stylistic improvements. It refers to the paper by T. Grund, C. Waloszek, and D. Helbing (2013) How natural selection can create both self- and other-regarding preferences, and networked minds. Sci. Rep. 3:1480, see http://www.nature.com/srep/2013/130319/srep01480/ full/srep01480.html.

In their computer simulations of human evolution, scientists at ETH Zurich find the emergence of the "homo socialis" with "other-regarding" preferences. The results explain some intriguing findings in experimental economics and call for a new economic theory of "networked minds".

Economics has a beautiful body of theory. But does it describe real markets? Doubts have emerged not only in the wake of the financial crisis, since financial crashes should not occur according to the then established theories. For ages, economic theory has been based on concepts such as efficient markets and the "homo economicus", i.e. the assumption of competitively optimizing individuals and firms. It was believed that any behavior deviating from this would create disadvantages and, hence, be eliminated by natural selection. But experimental evidence from behavioral economics show that, on average, people behave more fairness-oriented and other-regarding than expected. A new theory by scientists from ETH Zurich now explains why.

"We have simulated interactions of individuals facing social dilemma situations, where it would be favorable for everyone to cooperate, but non-cooperative behavior is tempting," explains Thomas Grund, one of the authors of the study. "Hence, cooperation tends to erode, which is bad for everyone." This may create tragedies of the commons such as over-fishing, environmental pollution, or tax evasion.

4.1   Evolution of "Friendliness"

Dirk Helbing of ETH Zurich, who coordinated the study, adds: "Compared to conventional models for the evolution of social cooperation, we have distinguished between the actual behavior—cooperation or not—and an inherited character trait, describing the degree of other-regarding preferences, which we call the friendliness." The actual behavior considers not only the own advantage ("payoff"), but also gives a weight to the payoff of the interaction partners depending on the individual friendliness. For the "homo economicus", the weight is zero. The friendliness spreads from one generation to the next according to natural selection. This is merely based on the own payoff, but mutations happen.

For most parameter combinations, the model predicts the evolution of a payoff-maximizing "homo economicus" with selfish preferences, as assumed by a great share of the economic literature. Very surprisingly, however, biological selection may create a "homo socialis" with other-regarding preferences, namely if offsprings tend to stay close to their parents. In such a case, clusters of friendly people, who are "conditionally cooperative", may evolve over time. If an unconditionally cooperative individual is born by chance, it may be exploited by everyone and not leave any offspring. However, if born in a favorable, conditionally cooperative environment, it may trigger cascade-like transitions to cooperative behavior, such that other-regarding behavior pays off. Consequently, a "homo socialis" spreads.

4.2 Networked Minds Create a Cooperative Human Species

"This has fundamental implications for the way, economic theories should look like," underlines Dirk Helbing. Most of today's economic knowledge is for the "homo economicus", but people wonder whether that theory really applies. A comparable body of work for the "homo socialis" still needs to be written.

"While the 'homo economicus' optimizes its utility independently, the 'homo socialis' puts himself or herself into the shoes of others to consider their interests as well," explains Grund, and Helbing adds: "This establishes something like 'networked minds'. Everyones decisions depend on the preferences of others." This becomes even more important in our networked world.

4.3 A Participatory Kind of Economy

How will this change our economy? Today, many customers doubt that they get the best service by people who are driven by their own profits and bonuses. "Our theory predicts that the level of other-regarding preferences is distributed broadly, from selfish to altruistic. Academic education in economics has largely promoted the selfish type. Perhaps, our economic thinking needs to fundamentally change, and our economy should be run by different kinds of people," suggests Grund. "The true capitalist has other-regarding preferences," adds Helbing, "as the 'homo socialis' earns much more payoff." This is, because the "homo socialis" manages to overcome the downwards spiral that tends to drive the "homo economicus" towards tragedies of the commons. The breakdown of trust and cooperation in the financial markets back in 2008 might be seen as good example.

"Social media will promote a new kind of participatory economy, in which competition goes hand in hand with cooperation," believes Helbing. Indeed, the digital economy's paradigm of the "prosumer" states that the Internet, social platforms, 3D printers and other developments will enable the co-producing consumer. "It will be hard to tell who is consumer and who is producer", says Christian Waloszek. "You might be both at the same time, and this creates a much more cooperative perspective."

5

A New Kind of Economy is Born – Social Decision-Makers Beat the "Homo Economicus"

This chapter was first published on October 8, 2013, in Today under the title "A new kind of economy is born", see http://www.todayonline.com/singapore/new-kind-economy-born, and is reproduced here with minor stylistic improvements. It refers to the Discussion Paper entitled Economics 2.0: The Natural step towards a self-regulating, participatory market society, see http://www.researchgate.net/profile/Dirk_Helbing/publication/236858 622_Economics_2.0_The_Natural_Step_towards_A_Self-Regulating_ Participatory_Market_Society/links/02e7e5246c4599ee9b000000.pdf.

The Internet and Social Media change our way of decision making. We are no longer the independent decision-makers we used to be. Instead, we have become networked minds, social decision-makers, more than ever before. This has several fundamental implications. First of all, our economic theories must change, and second, our economic institutions must be adapted to support the social decision-maker, the "homo socialis", rather than tailored to the perfect egoist, known as "homo economicus".

The financial, economic and public debt crisis has seriously damaged our trust in mainstream economic theory. Can it really offer an adequate description of economic reality? Laboratory experiments keep questioning one of the main pillars of economic theory, the "homo economicus". They show that the perfectly self-regarding decision-maker is not the rule, but rather the exception [1, 2]. And they show that markets, as they are organized today, are undermining ethical behavior [3].

Latest scientific results have shown that a "homo socialis" with other-regarding preferences will eventually result from the merciless forces of evolution, even if people optimize their utility, when offspring tend to stay close to their parents [4]. Another, independent study was recently summarized by the statement "evolution will punish you, if you're selfish and mean" [5]. Is this really true? And what implications would this have for our economic theory and institutions?

In fact, the success of the human species as compared to others results mainly from its social nature. There is much evidence that evolution has created different incentive systems, not just one: besides the desire to possess (in order to survive in times of crises), this includes sexual satisfaction (to ensure reproduction), curiosity and creativity (to explore opportunities and risks), emotional satisfaction (based on empathy), and social recognition (reputation, power). Already Adam Smith noted: "How ever selfish man may be supposed, there are evidently some principles in his nature, which interest him in the fortune of others, and render their happiness necessary to him, though he derives nothing from it."

The social nature of man has dramatic implications, both for economic theory and for the way we need to organize our economy. As we are more and more connected with others, the "homo economicus", i.e. the independent decision-maker and perfect egoist, is no longer an adequate representation or good approximation of human decision-makers. Reality has changed. We are applying an outdated theory, and that's what makes economic crises more severe.

5.1 Outdated Theory, Outdated Institutions

In fact, recent experimental results suggest that the majority of decision-makers are of the type of a "homo socialis" with equity- or equality-oriented fairness preferences [1, 6]. The "homo socialis" is characterized by two features: interdependent decision-making, which takes into account the impact on others, and conditional co-operativeness. Note that the "homo socialis" takes self-determined, free decisions. However, in contrast to what we have today, the principle is not to encourage everyone to rip off others, and then afterwards to give back some of the individual benefits to those in need through taxes or philanthropy—as we know, this cannot overcome "tragedies of the commons" (see below).

The "homo socialis" rather decides in a smarter way than the "homo economicus", recognizing that friendly and fair behavior can generate better outcomes for everybody than if everyone or every company is just thinking for their own concern. Interestingly, putting oneself into the shoes of others when taking decisions creates interdependent deci-sions, "networked minds". Such "networked minds" enable collective intelligence, i.e. they can take more intelligent and better decisions than a single mind can do.

However, other-regarding preferences are vulnerable to ex-ploitation by the "homo economicus". In a selfish environment, the "homo socialis" cannot thrive. In other words, if the settings are not right, the "homo socialis" behaves the same as the "homo economicus". That's probably why we haven't noticed its existence for a long time. Our theories and institutions were tailored to the "homo economicus", not to the "homo socialis".

In fact, some of today's institutions, such as homogeneous mar-kets with anonymous exchange, undermine cooperation in social dilemma situations, i.e. situations in which cooperation would be favorable for everyone, but non-cooperative behavior can provide additional benefits (see [7]: Fig. 2).

5.2 New Institutions for a Global Information Society

People have built public roads, parks and museums, schools, libraries, universities, and global markets. What would be suitable institutions for the twenty-first century? Reputation systems can transfer the success principles of social communities to our globalized society, the global village. Most people and companies care about reputation. Therefore, reputation systems could support socially oriented decision-making and cooperation, with better outcomes for everyone [8]. In fact, reputation systems spread on the Web 2.0 like wildfire. People rate products, sellers, news, everything, be it at amazon, ebay, or trip adviser. We have become a "like it" generation, because we listen to what our friends like.

Importantly, recommender systems should not narrow down socio-diversity, as this is the basis of happiness, innovation and societal resilience. We don't want to live in a filter bubble, where we don't get a good picture of the world anymore, as Eli Pariser has pointed out [9]. Therefore, reputation systems should be pluralistic, open, and user-centric. Pluralistic reputation systems are oriented at the values and quality criteria of individuals rather than recommending what a company's reputation filter thinks is best. Self-determination of the user is central. We must be able to use different filters, choose the filters ourselves, and modify them. The diverse filters would mine the ratings and comments that people leave on the Web, but also consider how much one trusts in certain information sources.

Reputation creates benefits for buyers and sellers. A recent study shows that good reputation allows sellers to take a higher price, while customers can expect a better service [10]. Reputation systems may also promote better quality as well as socially and environmentally friendly production. This could be a new approach to reach more sustainable production, based on self-regulation rather than enforcement by laws. One day, reputation systems may also be used to create a new kind of money. The value of "qualified money" would depend on it's reputation and thereby create incentives to invest in ways that increase a money unit's reputation. It might create a more adaptive financial system and help to mitigate the recurrent crises we have been facing for hundreds of years. But the details still have to be worked out.

5.3 Benefits of a Self-Regulating Economy

Reputation systems could overcome some of the unwanted side effects of anonymous exchange thanks to pseudonymous or personal interactions. Thereby, they could potentially counter "tragedies of the commons" such as global warming, environmental exploitation and degradation, overfishing,... —constituting some of our major unsolved global problems. We can encounter such kinds of "social dilemma problems" everywhere. So far, governments try to fix them with top-down regulations and punitive institutions. However, these are very expensive, and often quite ineffective. Basically all industrialized countries suffer from exploding debts. In many countries, we cannot pay for this much longer, we are at the limit. We need a new approach. As Albert Einstein pointed out: "We cannot solve our problems with the same kind of thinking that created them."

Institutions supporting the "homo socialis" such as suitably designed reputation systems would enable a self-regulation of socio-economic systems. But self-regulation does not mean that everyone can choose the rules he likes. It only works with an other-regarding element. The self-regulation rules must be able to achieve a balance between the interests of everyone affected by the externalities of a decision.

Other-regarding decisions can overcome the classical conflict between economic and social motives. Self-regulation could also overcome the struggle between the bottom-up organization of markets and the top-down regulation by politics. This would remove a lot of friction from our current system, making it much more efficient—in the same way as the transition from centrally planned economies to self-organized markets has often created huge efficiency gains.

This can be illustrated with an example from urban traffic management. Traffic control is a problem where not everybody's desires can be satisfied immediately and at the same time, like in economic systems. It is a so-called NP-hard optimization problem—the computational effort explodes with system size, as for many economic optimization problems, e.g. in production and logistics. The study compares three kinds of control: A centralized top-down regulation by a traffic center, the classical control approach, and two decentralized control approaches. The first one assumes that each intersection independently minimizes the waiting times of approaching vehicles, as a "homo economicus" would do. The second one decides in an other-regarding way: it interrupts the minimization of waiting times, when this is needed to avoid spill-over effects at neighboring intersections. Summarizing, the "homo economicus" approach works well up to a moderate utilization of intersections, but queue lengths get out of control long before the intersection capacity is reached. The bottom-up self-regulation based on the principle of the "homo socialis" approach beats both, the centralized top-down regulation and the bottom-up self-organization based on principles of the "homo economicus". Other-regarding behavior improves the coordination among neighboring intersections. It makes the principle of the "invisible hand" work, even at high utilizations.

5.4   Economics 2.0: Emergence of a Participatory Market Society

But will such a self-regulating system ever be implemented? In fact, this new, third kind of economy is already on its way. The Web 2.0, in particular reputation systems and social media are driving the transition towards an economy 2.0. We see already a strong trend towards decentralized, local production and personalized products, enabled by 3D printers, app stores, and other technologies.

Such developments will eventually create a participatory market society. "Prosumers", i.e. co-producing consumers, the new "makers" movement, and the sharing economy are some examples illustrating this. Just think of the success of Wikipedia, Open Streetmap or Github. Open Streetmap now provides the most up-to-date maps of the world, thanks to more than 1 million volunteers. This is just the beginning of a new era, where production and public engagement will more and more happen in a bottom up way through fluid "projects", where people can contribute as a leaders ("entrepreneurs") or participants. A new intellectual framework is emerging, and a creative and participatory era is ahead. The paradigm shift towards participatory bottom-up self-regulation may be bigger than the paradigm shift from a geocentric to a heliocentric worldview. If we build the right institutions for the information society of the twenty-first century, we will finally be able to mitigate some very old problems of humanity. "Tragedies of the commons" are just one of them. After so many centuries, they are still plaguing us, but this needn't be the case.

References

1. J. Henrich, R. Boyd, S. Bowles, C. Camerer, E. Fehr, H. Gintis, R. McElreath, In search of homo economicus: behavioral experiments in 15 small-scale societies. Am. Econ. Rev. **91**, 73–78 (2001)

2. R.O. Murphy, K.A. Ackermann, M.J.J. Handgraaf, Measuring social value orientation. Judgm. Decis. Mak. **6**(8), 771–781 (2011)

3. A. Falk, N. Szech, Morals and markets. Science **340**, 707–711 (2013)

4. T. Grund, C. Waloszek, D. Helbing, How natural selection can create both self-and other-regarding preferences, and networked minds. Sci. Rep. **3**, 1480 (2013), http://www.nature.com/srep/2013/130319/srep01480/full/srep01480.html

5. C. Adami, A. Hintze, Evolutionary instability of zerodeterminant strategies demonstrates that winning is not everything. Nat. Commun. **4**, 2193 (2013); Evolution will punish you, if you're selfish and mean. http://esciencenews.com/articles/2013/08/01/evolution.will.punish.you.if.youre.selfish.and.mean

6. R. Berger, H. Rauhut, S. Prade, D. Helbing, Bargaining over waiting time in ultimatum game experiments. Soc. Sci. Res. **41**, 372–379 (2012)

7. D. Helbing, Globally networked risks and how to respond. Nature **497**, 51–59 (2013)

8. M. Milinski, D. Semmann, H.J. Krambeck, Reputation helps solve the tragedy of the commons. Nature **415**, 424–426 (2002)

9. E. Pariser, Filter Bubble (Carl Hanser, Munich, 2012)

10. W. Przepiorka, Buyers pay for and sellers invest in a good reputation: more evidence from eBay. J. Soc-Econ. **42**, 31–42 (2013)

Further Reading

1. E.M. Johnson, Human nature and the moral economy, Scientific American Blog, http://blogs.scientificamerican.com/primate-diaries/2013/09/23/human-nature-and-the-moral-economy/?print=true

2. D. Helbing, Economics 2.0: The Natural step towards a self-regulating, participatory market society, Evolutionary and Institutional Economics Review (2013), see https://www.jstage.jst.go.jp/article/eier/10/1/10_3/_article. and watch the movie at https://www.youtube.com/watch?v=ZHYxMHm4t6U or http://www.youtube.com/watch?v=Ef2Ag_rwouo

6
Global Networks Must be Redesigned

This chapter was first published as press release of ETH Zurich on April 30, 2013, see http://www.alphagalileo.org/ViewItem.aspx? ItemId=130736&CultureCode=en, and is reproduced here with minor stylistic improvements. It features the publication "Globally networked risks and how to respond" which appeared in Nature 497, 51–59 (2013), see http://www.researchgate.net/ publication/236602842_Globally_networked_risks_and_ how_to_respond/file/60b7d52ada0b3d1494.pdf.

Todays strongly connected, global networks have produced highly interdependent systems that we have not been able to adequately understand and control. These systems are vulnerable to failure at all scales, posing serious threats to society, even when external shocks are absent. As the complexity and interaction strengths in our networked world increase, man-made systems can become unstable, creating uncontrollable situations even when decision-makers are well-skilled, have all data and technology at their hands, and do their best. To make these systems manageable, a fundamental redesign is needed. A Global Systems Science might create the required knowledge and paradigm shift in thinking.

6.1   Living in a Hyperconnected World

Our global networks have generated many benefits and new opportunities. However, they have also established highways for failure propagation, which can ultimately result in man-made disasters. For example, today's quick spreading of emerging epidemics is largely a result of global air traffic, with serious impacts on global health, social welfare, and economic systems.

Helbings publication illustrates how cascade effects and complex dynamics amplify the vulnerability of networked systems. For example, just a few long-distance connections can largely decrease our ability to mitigate the threats posed by global pandemics. Initially beneficial trends, such as globalization, increasing network densities, higher complexity, and an acceleration of institutional decision processes may ultimately push man-made or human-influenced systems towards systemic instability, Helbing finds. Systemic instability refers to a system, which will get out of control sooner or later, even if everybody involved is well skilled, highly motivated and behaving properly. Crowd disasters are shocking examples illustrating that many deaths may occur even when everybody tries hard not to hurt anyone.

6.2   Our Intuition of Systemic Risks is Misleading

Networking system components that are well-behaved in separation may create counter-intuitive emergent system behaviors, which are not well-behaved at all. For example, cooperative behavior might unexpectedly break down as the connectivity of interaction partners grows. "Applying this to the global network of banks, this might actually have caused the financial meltdown in 2008," believes Helbing.

Globally networked risks are difficult to identify, map and understand, since there are often no evident, unique cause-effect relationships. Failure rates may change depending on the random path taken by the system, with the consequence of increasings risks as cascade failures progress, thereby decreasing the capacity of the system to recover. "In certain cases, cascade effects might reach any size, and the damage might be practically unbounded," says Helbing. "This is quite disturbing and hard to imagine." All of these features make strongly coupled, complex systems difficult to predict and control, such that our attempts to manage them go astray.

"Take the financial system," says Helbing. "The financial crisis hit regulators by surprise." But back in 2003, the legendary investor Warren Buffet warned of mega-catastrophic risks created by large-scale investments into financial derivatives. It took 5 years until the "investment time bomb" exploded, causing losses of trillions of dollars to our economy. "The financial architecture is not properly designed," concludes Helbing. "The system lacks breaking points, as we have them in our electrical system." This allows local problems to spread globally, thereby reaching catastrophic dimensions.

6.3   A Global Ticking Time Bomb?

Have we unintentionally created a global time bomb? If so, what kinds of global catastrophic scenarios might humans face in complex societies? A collapse of the world economy or of our information and communication systems? Global pandemics? Unsustainable growth or environmental change? A global food or energy crisis? A cultural clash or global-scale conflict? Or will we face a combination of these contagious phenomena a scenario that the World Economic Forum calls the "perfect storm"?

"While analyzing such global risks," says Helbing, "one must bear in mind that the propagation speed of destructive cascade effects might be slow, but nevertheless hard to stop. It is time to recognize that crowd disasters, conflicts, revolutions, wars, and financial crises are the undesired result of operating socio-economic systems in the wrong parameter range, where systems are unstable." In the past, these social problems seemed to be puzzling, unrelated, and almost "God-given" phenomena one had to live with. Nowadays, thanks to new complexity science models and large-scale data sets ("Big Data"), one can analyze and understand the underlying mechanisms, which let complex systems get out of control.

Disasters should not be considered "bad luck". They are a result of inappropriate interactions and institutional settings, caused by humans. Even worse, they are often the consequence of a flawed understanding of counter-intuitive system behaviors. "For example, it is surprising that we didnt have sufficient precautions against a financial crisis and well-elaborated contingency plans," states Helbing. "Perhaps, this is because there should not be any bubbles and crashes according to the predominant theoretical paradigm of efficient markets." Conventional thinking can cause fateful decisions and the repetition of previous mistakes. "In other words: While we want to do the right thing, we often do wrong things," concludes Helbing. This obviously calls for a paradigm shift in our thinking. "For example, we may sanction deviations from social norms to promote social order, but may trigger conflict instead. Or we may increase security measures, but get more terrorism. Or we may try to promote innovation, but suffer economic decline, because innovation requires diversity more than homogenization."

6.4 Global Networks Must be Redesigned

Helbings publication explores why todays risk analysis falls short. "Predictability and controllability are design issues," stresses Helbing. "And uncertainty, which means the impossibility to determine the likelihood and expected size of damage, is often man-made." Many systems could be better managed with real-time data. These would allow one to avoid delayed response and to enhance the transparency, understanding, and adaptive control of systems. However, even all the data in the world cannot compensate for ill-designed systems such as the current financial system. Such systems will sooner or later get out of control, causing catastrophic man-made failure. Therefore, a re-design of such systems is urgently needed.

Helbings Nature paper on "Globally Networked Risks" also calls attention to strategies that make systems more resilient, i.e. able to recover from shocks. For example, setting up backup systems (e.g. a parallel financial system), limiting the system size and connectivity, building in breaking points to stop cascade effects, or reducing complexity may be used to improve resilience. In the case of financial systems, there is still much work to be done to fully incorporate these principles.

Contemporary information and communication technologies (ICT) are also far from being failure-proof. They are based on principles that are 30 or more years old and not designed for todays use. The explosion of cyber risks is a logical consequence. This includes threats to individuals (such as privacy intrusion, identity theft, or manipulation through personalized information), to companies (such as cybercrime), and to societies (such as cyberwar or totalitarian control). To counter this, Helbing recommends an entirely new ICT architecture inspired by principles of decentralized self-organization as observed in immune systems, ecology, and social systems.

6.5   Coming Era of Social Innovation

Socio-inspired technologies built on decentralized mechanisms that create reputation, trust, norms or culture will be able to generate enormous value. "Facebook, based on the simple principle of social networking, is worth more than 50 billion dollars," Helbing reminds us. "ICT systems are now becoming artificial social systems. Computers already perform the great majority of financial transactions, which humans carried out in the past." But if we do not understand socially interactive systems well, coordination failures, breakdowns of cooperation, conflict, cyber-crime or cyber-war may result.

Therefore, a better understanding of the success principles of societies is urgently needed. "For example, when systems become too complex, they cannot be effectively managed top-down" explains Helbing. "Guided self-organization is a promising alternative to manage complex dynamical systems bottom-up, in a decentralized way." The underlying idea is to exploit, rather than fight, the inherent tendency of complex systems to self-organize and thereby create a robust, ordered state. For this, it is important to have the right kinds of interactions, adaptive feedback mechanisms, and institutional settings, i.e. to establish proper "rules of the game". The paper offers the example of an intriguing "self-control" principle, where traffic lights are controlled bottom-up by the vehicle flows rather than top-down by a traffic center.

6.6 Creating and Protecting Social Capital

It is important to recognize that many twenty-first century challenges such as the response to global warming, energy and food problems have a social component and cannot be solved by technology alone. The key to generating solutions is a Global Systems Science (GSS) that brings together crucial knowledge from the natural, engineering and social sciences. The goal of this new science is to gain an understanding of global systems and to make "systems science" relevant to global problems. In particular, this will require the combination of the Earth Systems Sciences with the study of behavioral aspects and social factors.

"One mans disaster is another mans opportunity. Therefore, many problems can only be successfully addressed with transparency, accountability, awareness, and collective responsibility," underlines Helbing. "For example, social capital is important for economic value generation, social well-being and societal resilience, but it may be damaged or exploited, like our environment," explains Helbing. "Humans must learn how to quantify and protect social capital. A warning example is the loss of trillions of dollars in the stock markets during the financial crisis." This crisis was largely caused by a loss of trust.

"It is important to stress that risk insurances today do not consider damage to social capital," Helbing continues. However, it is known that large-scale disasters have a disproportionate public impact, in part because they destroy social capital. As we neglect social capital in risk assessments, we are taking excessive risks.

7

Big Data—A Powerful New Resource for the Twenty-first Century

This chapter is a translation of an introductory article on "Big Data—Zauberstab und Rohstoff des 21. Jahrhunderts" published in Die Volkswirtschaft—Das Magazin für Wirtschaftspolitik (5/2014), see http://www.dievolkswirtschaft.ch/files/editions/201405/pdf/04_Helbing_DE.pdf. It reproduces the FuturICT blog of September 26, 2014, see http://futurict.blogspot.ch/2014/09/big-data-powerful-new-resource-for-21st.html, with minor improvements.

Information and communication technology (ICT) is the economic sector that is developing most rapidly in the USA and Asia and generates the greatest value added per employee. Big Data—the algorithmic discovery of hidden treasures in large data sets—creates new economic value. The development is increasingly understood as a new technological revolution. Europe could establish itself as Open Data pioneer and turn into a leading place in the area of information technologies. When the social media portal WhatsApp with its 450 million users was recently sold to Facebook for $ 19 billion—almost half a billion dollars was made per employee. "Big Data" is changing our world. The term, coined more than 15 years ago, means data sets so big that one can no longer cope with them with standard computational methods. Big Data is increasingly referred to as the oil of the twenty-first century. To benefit from it, we must learn to "drill" and "refine" data, i.e. to transform them into useful information and knowledge. The global data volume doubles every 12 months. Therefore, in just two years, we produce as much data as in the entire history of humankind.

Tremendous amounts of data have been created by four technological innovations:

- the Internet, which enables our global communication,
- the World Wide Web, a network of globally accessible websites that evolved after the invention of hypertext protocol (HTTP) at CERN in Geneva,
- the emergence of social media such as Facebook, Google+, Whatsup, or Twitter, which have created social communication networks, and
- the emergence of the "Internet of Things", which also allows sensors and machines to connect to the Internet. Soon there will be more machines than human users in the Internet.

7.1 Data Sets Bigger than the Largest Library

Meanwhile, the data sets collected by companies such as ebay, Walmart or Facebook, reach the size of petabytes (1 million billion bytes)-100 times the information content of the largest library in the world: the U.S. Library of Congress. The mining of Big Data opens up entirely new possibilities for process optimization, the identification of interdependencies, and decision support. However, Big Data also comes with new challenges, which are often characterized by four criteria:

- *volume:* the file sizes and number of records are huge,
- *velocity:* the data evaluation has often to be done in real-time,

- *variety:* the data are often very heterogeneous and unstructured,
- *veracity:* the data are probably incomplete, not representative, and contain errors.

Therefore, one had to develop completely new algorithms: new computational methods. Because it is inefficient for Big Data processing to load all relevant data into a shared memory, the processing must take place locally, where the data reside, potentially on thousands of computers. This is accomplished with massively parallel computing approaches such as *MapReduce* or *Hadoop*. Big Data algorithms detect interesting interdependencies in the data ("correlations"), which may be of commercial value, for example, between weather and consumption or between health and credit risks. Today, even the prosecution of crime and terrorism is based on the analysis of large amounts of behavioral data.

7.2   What Do Applications Look Like?

Big Data applications are spreading like wildfire. They facilitate personalized offers, services and products. One of the greatest successes of Big Data is automatic speech recognition and processing. Apple's Siri understands you when asking for a Spanish restaurant, and Google Maps can lead you there. Google translate interprets foreign languages by comparing them with a huge collection of translated texts. IBM's Watson computer even understands human language. It can not only beat experienced quiz show players, but even take care of customer hotlines—often better than humans. IBM has just decided to invest $ 1 billion to further develop and commercialize the system.

Of course, Big Data play an important role in the financial sector. Approximately 70 % of all financial market transactions are now made by automated trading algorithms. In just a day, the entire money supply of the world is traded. So much money also attracts organized crime. Therefore, financial transactions are scanned by Big Data algorithms for abnormalities to detect suspicious activities. The company blackrock uses a similar software, called "Aladdin", to successfully speculate with funds amounting approximately to the gross domestic product (GDP) of Europe.

To get an overview of the ICT trends, it is worthwhile to look at Google with its more than 50 software platforms. The company invests nearly $ 6 billion in research and development annually. Within just one year, Google has introduced self-driving cars, invested heavily in robotics, and started a Google Brain project, which intends to add intelligence to the Internet. By purchasing Nest Labs, Google has also invested $ 3.2 billion in the "Internet of Things". Furthermore, Google X has been reported to work on about 100 secret projects.

7.3   The Potentials Are Great...

No country today can afford to ignore the potentials of Big Data. The additional economic potential of Open Data alone—i.e. of data sets that are made available to everyone—is estimated by McKinsey to be 3000–5000 billion dollars globally each year [1]. This can benefit almost all sectors of society. For example, energy production and consumption can be better matched with "smart metering", and energy peaks can be avoided.[1] Resources can be managed more efficiently and the environment protected better. Risks can be better recognized and avoided, thereby reducing unintended consequences of decisions and identifying opportunities that would otherwise have been missed. Medicine can be better

---

[1] More generally, new information and communication technologies allow us to build "smart cities".

adapted to the patients, and disease prevention may become more important than curing diseases.

7.4 ... but also the Implicit Risks

Like all technologies, Big Data also imply risks. The security of digital communication has been undermined. Cyber crime, including data, identity and financial theft, quickly spread on ever greater dimensions. Critical infrastructures such as energy, financial and communication systems are threatened by cyber attacks. They could, in principle, be made dysfunctional for an extended period of time.

Moreover, while common Big Data algorithms are used to reveal optimization potentials, their results may be unreliable or may not reflect causal relationships. Therefore, a naive application of Big Data algorithms can easily lead to wrong conclusions. The error rate in classification problems (e.g. the distinction between "good" and "bad" risks) is often relevant. Issues such as wrong decisions or discrimination must be seriously considered. Therefore, one must find effective procedures for quality control. In this connection, universities will likely play an important role. One must also find effective mechanisms to protect privacy and the right of informational self-determination, for example, by applying the Personal Data Purse concept [2].

7.5 The Digital Revolution Creates an Urgency to Act

Information and communication technologies are going to change most of our traditional institutions: our educational system (personalized learning), science (Data Science), mobility (self-driving cars), the transport of goods (drones), consumption (see amazon and ebay), production (3D printers), the health system (personalized medicine), politics (more transparency), and the entire economy (with co-producing consumers, so-called prosumers). Banks are losing more and more ground to algorithmic trading, Bitcoins, Paypal and Google Wallet. Moreover, a substantial part of the insurance business is now taking place in financial products such as credit default swaps. For the economic and social transformation into a "digital society", we may perhaps just have 20 years. This is an extremely short time period, considering that the planning and construction of a road often requires 30 years or more.

The above implies an urgent need for action on the technological, legal and socio-economic level. Already some years ago, the United States started a Big Data research initiative amounting to 200 million dollars; on top of this, further substantial investments were made. In Europe, the FuturICT project has developed concepts for the digital society within the context of the EU flagship competition. Other countries have already started to implement this concept. Japan, for example, has recently launched a $ 100 million 10-year project at the Tokyo Institute of Technology. Besides, numerous further projects exist, particularly in the military and security sector, which often have a multiple of the above-mentioned budgets.

7.6   Europe can Become a Motor of Innovation for the Digital Era

Europe can benefit a lot from the digital age. But it is not sufficient to reinvent and build already existing technologies in Europe. We must come up with new inventions, which will shape the digital age. The World Wide Web was once invented in Europe. The largest civil Big Data competence in the world exists at CERN. Nevertheless, the USA and Asian countries so far have the lead in commercializing Big Data. With the NSA scandal, the spreading of wirelessly communicating sensors and the "Internet of Things", however, a new opportunity is emerging.

With a targeted support of ICT activities at its universities, Europe could take a lead in research and development. Switzerland, in particular, has academically excelled with the coordination of three out of six finalists of the EU flagship competition. At the moment, however, there is only a focus on the digital modeling of the human brain and robotics. From 2017 onwards, the ETH domain plans to increasingly invest into the area of *Data Science*, the emerging research field centered around the scientific analysis of data. However, in view of the fast development of the ICT area, the huge economic potential, and also the transformative power of these technologies, a prioritized, broad and substantial financial support is a matter of national interest.

With its democratic values, its legal framework, and its ICT engagement, including industry 4.0-related business, Europe might become an innovation motor for the digital age.

References

1. McKinsey & Company, Open data: unlocking innovation and performance with liquid information. http://www.mckinsey.com/insights/business_technology/open_data_unlocking_innovation_and_performance_with_liquid_information (2013). Accessed 28 April 2014
2. Y.-A. de Montjoye, E. Shmueli, S.S. Wang, A.S. Pentland, openPDS: protecting the privacy of metadata through SafeAnswers. PLoS ONE **9**(7), e98790. http://www.plosone.org/article/info%3Adoi%2F10.1371%2Fjournal.pone.0098790; see also http://newsoffice.mit.edu/2014/own-your-own-data-0709, http://www.taz.de/!131892/, and http://www.weforum.org/reports/personal-data-emergence-new-asset-class (2014). Accessed 28 April 2014

8

Google as God? Opportunities and Risks of the Information Age

This chapter is the English translation of an article entitled Google als Gott?, which appeared in the Neue Zürcher Zeitung on March 20, 2013, see http://www.nzz.ch/aktuell/wirtschaft/wirtschaftsnachrichten/google-als-gott-1.18049950; for the first appearance of the English version see http://futurict.blogspot.de/2013/03/google-as-god-opportunities-and-risks.html, which is reproduced here with minor stylistic improvements.

If God did not exist—people would invent one! The development of human civilization requires mechanisms promoting cooperation and social order. One of these mechanisms is based on the idea that everything we do is seen and judged by God—bad deeds will be punished, while good ones will be rewarded. The Information Age has now fueled the dream that God-like omniscience and omnipotence can be created by man. This essay discusses the implications.

8.1 Introduction

You're already a walking sensor platform... You are aware of the fact that somebody can know where you are at all times because you carry a mobile device, even if that mobile device is turned off. You know this, I hope? Yes? Well, you should... Since you can't connect dots you don't have, it drives us into a mode of, we fundamentally try to collect everything and hang on to it forever... It is really very nearly within

our grasp to be able to compute on all human generated information.
Ira "Gus" Hunt, CIA Chief Technology Officer

For many decades, the processing power of computer chips has increased exponentially—as predicted by "Moore's Law." Storage capacity is growing even faster. We are now entering a phase of the "Internet of Things", where computer chips and measurement sensors will soon be scattered everywhere producing huge amounts of data ("Big Data"). Cell phones, computers and factories, but also our coffee machines, fridges, shoes and clothes, among others, are getting more and more connected.

8.2   Gold Rush for the Twenty-first Century Oil

This huge amount of data, including credit card transactions, communication with friends and colleagues, mobility data and more is already celebrated as the "Oil of the Twenty-first Century". The gold rush to exploit this valuable resource is just starting. The more data are generated, stored and interpreted, the easier is it for companies and secret services to know us better than our friends and families do. For example, the company "Recorded Future"—apparently a joint initiative between Google and the CIA—seems to investigate people's social networks and mobility profiles. Furthermore, credit card companies analyze "consumers' genes"—the factors that determine our consumer behavior.

Our individual motivations are analyzed in order to understand our decisions and influence our behavior through personalized search, individualized advertisements, and recommendations or decisions of our Facebook friends. But how many of these "friends" are trustable, how many of them are paid to influence us, and how many are software robots?

8.3   Humans Controlled by Computers?

Today, computers autonomously perform the majority of financial transactions. They decide how much we have to pay for our loans or insurances, based on our behavioral data and on those of our friends, neighbors and colleagues. People are increasingly discriminated by obscure "machine learning" algorithms, which are neither transparent nor have to meet particular quality standards. People classified as dangerous are now eliminated by drones, without a chance to prove their innocence, while some countries are discussing robots' rights. Soon, Google will drive our cars. And in 10 years, supercomputers will exceed the performance of human brains.

8.4   Is Privacy Still Needed?

What will the role of privacy be in such an information society? Some companies are already trying to turn privacy into a marketable commodity. This is done by first taking away our privacy and then selling it back to us. The company Acxiom, for example, is said to sell detailed data about more than 500 million people. Would it be possible to know beforehand whether the data will be used for good or bad? Many will pay to have their personal data removed from the Internet and commercial databases. And where data removal is not possible, fake identities and mobility profiles will be offered for sale, to obfuscate our traces.

8.5   Information Overload

"Big Data" do not necessarily mean that we will be able to perceive the world more accurately. Rather, we will have to pay for "digital eyewear" that allows us to keep an overview in the data deluge. Those not willing to pay (possibly also with personal data) will be blinded by an information overload. Already today, we cannot assess the quality of the answers we get online. As the way in which the underlying data are processed remains hidden to the user, it is hard to know how much we are being manipulated by web services and social media. But given the huge economic potential, it is pretty clear that manipulation is happening.

8.6   The Knowledge-Is-Power Society

The statement "knowledge is power" seems to imply that "omniscience is omnipotence"—a tempting idea indeed. Therefore, whoever collects all the data in the world, such as the National Security Agency (NSA) in the United States, might hope to become almighty, especially if equipped with suitable manipulation tools. By knowing everything about us, one can always find a weak spot. Even CIA director General David Petraeus was not immune to this risk. He became the victim of a love affair irrelevant to his duty.

The developments outlined above are not fantasy—they are already taking place behind the scenes or are just around the corner. Yet, our society and legal systems are not well prepared for this.

8.7   A New World Order Based on Information?

Some people may see information and technologies as new tools to create social order. Why should one object to a computer or company or government taking decisions for us, as long as they act in our interest? But who would decide how to use these tools?

Can the concept of a 'caring state' or a 'benevolent dictator' really work? In other words, can supercomputers enabled by Big Data take the best decisions for us?

This has always failed in the past, and will be unsuccessful in the future. Not only do many systems fail under asymmetric information (if some stakeholders are very well informed and others very badly). The performance of all computers in the world will never be sufficient to optimize our economy and society in real time. Supercomputers cannot even optimize the traffic lights of a big city in real time. This is because the computational effort explodes with the size and complexity of the system. Just a very simple society could be optimized top down, but who would want to live in it?

8.8 Privacy and Socio-Diversity Need Protection

The aforementioned "omniscient almighty society" cannot work. If we all did what is right according to a super-intelligent institution—it would be as if children always did what their parents are asking for. Then they would never learn to take autonomous decisions, and to go their own way. Privacy is a necessary ingredient for the development of individual responsibility and for society. It should not to be understood as a concession to the citizens.

"Private" and "public" are two sides of the same coin, which could not exist without each other. People can only adjust to the thousands of normative public expectations every day, if there is a private, protected space where they can be free and relax. Privacy reduces mutual interference to a degree that allows us to "live and let live". If we knew what others secretly think about us, we would have far more conflicts.

The importance of unobserved opinion formation is demonstrated by the crucial role of anonymous votes in democracies. Would we only adjust ourselves to expectations of others, many new ideas would not emerge or spread. Social diversity would decrease, and thus the ability of our society to adapt. Innovation requires the protection of minorities and new ideas. It is an engine of our economy. Social diversity also promotes happiness, social well-being, and the ability of our society to recover from shocks ("social resilience").

Social diversity must be protected just as biodiversity. Today, however, the Internet recommends books, music, movies to us, and it even suggests how to think about politics and other people. This undermines the "wisdom of crowds" and collective intelligence. Why should a company decide what is good for us? Why can't we determine the recommendation algorithms ourselves? Why don't we get access to relevant data?

8.9   An Alternative Vision of the Information Age

In an increasingly unstable world, surveillance, combined with the manipulation or suppression of undesired behaviors, is not a sustainable solution. But is there an alternative to the omniscient almighty state that matches our ethical values? An alternative that can create cultural and economic prosperity? Yes there is, indeed!

Our society and economy are currently undergoing a fundamental transformation. Global networking creates increasing complexity and instability that cannot be properly managed by planning, optimization and top-down control. A flexible adaptation to local needs works better for complex, variable systems. This means that managing complexity requires a stronger bottom-up component.

In the economy and the organization of the Internet, decentralized self-organization principles have always played a big role. Now they have also spread to intelligent energy networks ("smart grids") and traffic control. One day, societal decision-making and economic production processes will also be run in a more participatory way to better manage the increasing complexity. This seems to be the natural course of history. A growing desire of citizens to participate in social, political and economic affairs is already found in many parts of the world.

8.10 The Democratic, Participatory Market Society

In connection with a participatory economy, one often speaks of "prosumers," i.e. co-producing consumers. Advanced collaboration platforms will allow anyone to set up projects with others to create their own products, for example with 3D printers. Thus, classical companies and political parties and institutions might increasingly be replaced by project-based initiatives—a form of organization that I would like to call "democratic, participatory market society."

To ensure that the participatory market society will work well and create jobs on a large scale, the right decisions will have to be made. For example, it seems essential that the information systems of the future will be open, transparent and participatory. This requires that we create a participatory information and innovation ecosystem, i.e. to make large amounts of data accessible to everyone.

8.11   The Benefit of Opening Data to All

The great advantage of information is that it is (re)producible in a cheap and almost unlimited way, so that the eternal struggle for limited resources might be overcome. It is important that we take advantage of this and open the door to an age of creativity rather than limiting access to information, thereby creating artificial scarcity again. Today, many companies collect data, but lack access to other important data. The situation is as if everyone owned a few words of our language, but had to pay for the use of all the other words. It is pretty clear that, under such conditions, we could not fully capitalize on our communicative potentials.

To overcome this dissatisfactory data exchange situation and achieve "digital literacy," one could decide to open up data to all. Remember that in the past most countries decided to turn the privilege of reading and writing into a public good by establishing public schools. This step boosted the development of modern societies. Similarly, "Open Data" could boost the development of the information society, but the producers of data must be adequately compensated.

8.12   A New Paradigm to Manage Complexity

Access to data is essential for the successful management of complex dynamical systems, as it requires three elements: (i) proper systems design to support predictability and controllability, (ii) probabilistic short-term forecasts of the system dynamics, which need plenty of reliable real-time data, and (iii) suitable adaptive mechanisms ("feedback loops") that support the desired system behavior.

Managing complexity should build on the natural tendency of complex dynamical systems to self-organize. To enable self-organization, it is crucial to find the right institutional settings and suitable "rules of the game", while avoiding too much top-down control. Then, complex systems can essentially regulate themselves.

One must be aware, however, that complex systems often behave in counterintuitive ways. Hence, it is easy to choose the wrong rules, thereby ending up with suboptimal results, unwanted side effects, or unstable system behaviors that can lead to manmade disasters. The financial system, which went out of control, might serve as a warning. These problems have traditionally been managed by top-down regulation, which is usually inefficient and expensive.

8.13   Loss of Control due to a Wrong Way of Thinking

Whether a system can't be adequately managed or is self-organizing according to our intentions is a matter of systems design. If the system is designed in the wrong way, then it will get out of control sooner or later, even if all actors involved are highly trained, well equipped and highly motivated to do the right things. "Phantom traffic jams" and "crowd disasters" are examples of unwanted situations occurring despite all efforts to prevent them. Likewise, financial crises, conflicts and wars can be unintended consequences of systemic instabilities. Even today, we are still not immune to them.

Therefore, we need a deeper understanding of our technosocio-economic-ecological systems and their interdependencies. Appropriate institutions and rules for our highly networked world must still be found. The information age is revolutionizing our economy and society in a dramatic way. If we do not pay sufficient attention to these developments, we will suffer the fate of driving a car too fast on a foggy day.

8.14 Decisions Needed to Use Opportunities and Avoid Risks

To meet the challenges and benefit from the great opportunities of the twenty-first century, a Global Systems Science needs to be established in order to fill the current knowledge gaps. It aims to generate new insights allowing politics, economy and society to take better informed, more successful decisions. This could help us to use the opportunities provided by the information age and minimize its risks. We must be aware that everything is possible—ranging from a Big Brother society to a participatory economy and society. The choice is ours!

Further Reading

- **Difficulty to anonymize data:**
1. Researchers reverse Netflix anonymization, see http://www. securityfocus.com/news/11497
2. Unique in the crowd: The privacy bounds of human mobility, see http:// www.nature.com/srep/2013/130325/srep01376/ full/srep01376.html
- **Danger of surveillance society**
3. Big data is opening doors, but maybe too many http://www. nytimes.com/2013/03/24/technology/big-data-and-a-renewed -debate-over-privacy.html?ref=stevelohr&_r=2&

4. Future planet future of surveillance, see http://www.inter-national.to/index.php?option=com_content&view=category&id=94&layout=blog&Itemid=104

5. CIA and FBI strategies to mine personal data, see http://www.businessinsider.com/cia-presentation-on-big-data-2013-3?op=1, http://gigaom.com/2013/03/20/even-the-cia-is-struggling-to-deal-with-the-volume-of-real-time-social-data/2/, http://www.slate.com/blogs/future_tense/2013/03/26/andrew_weissmann_fbi_wants_real_time_gmail_dropbox_spying_power.html

- **New deal on data, how to consider consumer interests:**

6. US Consumer Privacy Bill of Rights, see http://www.white-house.gov/sites/default/files/privacy-final.pdf

7. Personal data: The emergence of a new asset class, see http://www.weforum.org/reports/personal-data-emergence-new-asset-class

8. HP software allowing personalized advertisement without revealing personal data to companies, contact: Prof. Dr. Bernardo Huberman <huberman@hpl.hp.com>

- **FuturICT initiative (http://www.futurict.eu):**

9. FuturICT The road towards ethical ICT, see http://link.springer.com/article/10.1140%2Fepjst%2Fe2012-01691-2#page-1

10. From social data mining to forecasting socio-economic crises, see http://epjst.epj.org/index.php?option=com_article&access=standard&Itemid=129&url=/articles/epjst/abs/2011/04/epjst195002/epjst195002.html

9

From Technology-Driven Society to Socially Oriented Technology: The Future of Information Society—Alternatives to Surveillance

This chapter is the English translation of an article entitled "Sozial orientierte Technologie", which appeared in the Neue Zürcher Zeitung on August 19, 2013, see http://www.nzz.ch/meinung/ debatte/sozial-orientierte-technologie-1.18135003; for the first appearance of the English version see http://futurict.blogspot.de/2013/ 07/from-technology-driven-society-to.html, which is reproduced here with minor stylistic improvements. The appendix appeared for the first time in the FuturICT blog http://futurict.blogspot.de/2013/ 06/why-mass-surveillance-does-not-work.html

Our society is changing. Almost nothing these days works without a computer chip. Computing power doubles every 18 months, and in 10 years it will probably exceed the capabilities of a human brain. Computers perform approximately 70 % of all financial transactions today and IBM's Watson now seems to give better customer advise than some human telephone hotlines. What does this imply for our future society?

The forthcoming economic and social transformation might be more fundamental than the one resulting from the invention of the steam engine. Meanwhile, the storage capacity of data grows even faster than the computational capacity. Within a few years, we will generate more data than in the entire history of humankind. The "Internet of Things" will soon network trillions of sensors together—fridges, coffee machines, electric toothbrushes and even our clothes. Vast amounts of data will be collected. Already, Big Data is being heralded as the oil of the twenty-first century.

But this situation will also make us vulnerable. Exploding cyber-crime, economic crises and social protests show that our hyper-connected world is destabilizing. However, is a Surveillance Society the right answer? When all our Internet queries are stored, when our purchases and social contacts are evaluated, when our emails and files are scanned for search terms, and when countless innocent citizens are classified as potential future terrorists, we must ask: Where will this lead to? And where will it end?

Will surveillance lead to self-censorship and discrimination against intellectuals and minorities, even though innovation and creative thinkers are bitterly needed for our economy and society to do well in our changing world? Will free human expression eventually be curtailed by data mining machines analyzing our digital trails? What are the consequences, say if even the Swiss banks and the U.S. government can no longer protect their secrets, or if our health and other sensitive data is sold on? Or if politically and commercially sensitive strategies can be monitored in real time? What if insider knowledge can be used to undermine fair competition and justice?

The recent allegations that information agencies of various states snoop secretly into the activities of millions of ordinary people has alarmed citizens and companies alike. The moral outrage in response to the surveillance activity has made it clear that it is not a technology-driven society that we need, but instead, a socially-oriented technology, as outlined below. We must recognize that technology without consideration of ethical issues, or without transparency and public discussions can lead us astray. Therefore a new approach to personal data and its uses is required so that we can safely benefit from the many new economic and social opportunities that it can provide.

First, we need a public ethical debate on the concepts of privacy and ownership of data, even more urgently than in bioethics. Important questions that we have to ask are: How do we create opportunities arising in the information age for all, but yet still manage the downside risks and challenges - from cyber-crime to the erosion of trust and democratic rights? Do we really need so much security that we must be afraid of data mining algorithms flagging the activities of millions of ordinary people as suspicious? And what kinds of new institutions would we need in the twenty-first century?

In the past we have built public roads, parks and museums, schools, libraries and universities. Now, more than ever, we need strategies that protect us against the misuse of data, and that are intended to create transparency and trust. These strategies must place citizen benefits and rights of self-determination at the very core. In addition, we must develop new institutions to provide oversight and control of the new challenges brought on by the data revolution. Here are some specific institutional proposals:

Self-Determined Use of Personal Data: Already some time ago, the World Economic Forum (WEF) called for a "New Deal on Data" (http://www.weforum.org/pdf/ gitr/2009/gitr09fullreport. pdf). It stated that the sustainable use of the economic opportunities of personal data requires a fair balance between economic, governmental and individual interests. A solution would be to return control over personal data to the respective individuals, i.e. give people ownership of their data: the right to possess, access, use and dispose. In addition, individuals should be able to participate in their economic profits. This would require new data protocols and the support of legislation.

Trusted Information Exchange   As the vulnerability of existing systems and the proliferation of cyber-crime indicates, a new network architecture is urgently needed. The handling of sensitive data requires secure encryption, anonymisation and protected pseudonyms, decentralized storage, open software codes and transparency on the use of data, correction possibilities, mechanisms of forgetting, and a protective "digital immune system."

Credibility Mechanisms   Social mechanisms such as reputation, as seen in the evaluation of information and information sources on the internet, can play a central role in reducing abuse. But remember that the wisdom of crowds only works if individual decisions are not manipulated. Therefore, to be effective, individuals must be given control over the recommendation mechanisms, data filtering and search routines they use, such that they can take decisions based on their own values and quality criteria.

Participatory Platforms   All over the world people desire increased participation, from consumption to production processes. Now, modern technology allows for the direct social, economic, and political participation of engaged individuals. A basic democracy approach as in Switzerland, where people can decide themselves about many laws, not just political representatives, would be feasible on much larger scales. We also witness an economic trend towards local production, ranging from solar panels to 3D printers. It can be become a good complement of mass production.

Open Data   The innovation ecosystem needs open data and open standards to flourish. Open data enable the rapid creation of new products, which stimulates further products and services. Information is the best catalyst for innovation. Of course, data providers should be adequately compensated, and not all data would have to be open.

Innovation Accelerator   To keep pace with our changing world, we need to reinvent the innovation process itself. A participatory innovation process would allow ideas to be implemented faster and external expertise to be integrated more readily. Information is an extraordinary resource: it does not diminish when shared, and it can be infinitely reproduced. Why shouldn't we use this opportunity?

Social Capital   Information systems can support diverse types of social capital such as trust, reputation, and cooperation. Based on social network interactions, they are the foundation of a flourishing economy and society. So, let's create new value!

Social Technologies   Finally, we must learn to build information systems that are compatible with our individual, social and cultural values. We need to design systems that respect the privacy of citizens and prevent fear and discrimination, while promoting tolerance, trust, and fairness. What solutions can we offer users to ensure that information systems are not misused for unjustified monitoring and manipulation? For a well-functioning society, socio-diversity (pluralism) must be protected as much as biodiversity. Both determine the potential for innovation.

These are just some examples of the promising ways in which we could use the Internet of the future. Among all these, a surveillance society is probably the worst of all uses of information technology. A safe and sustainable information society has to be built on reputation, transparency and trust, not mass surveillance. If we can no longer trust our phones, computers or the Internet, we will either switch off our equipment or start to behave like agents of a secret service: revealing as little information as possible, encrypting data, creating multiple identities, laying false traces.

Such behaviour would create little benefits for ordinary citizens, besides protection, but might help criminals to hide. It would be a pity if we failed to use the opportunities afforded by the information age, just because we did not think hard or far enough about the technological and legal frameworks and institutions needed.

The information age is now at a crossroad. It may eventually lead us to a totalitarian surveillance state, or we can use it towards a creative, participatory society. It is our decision, and we should not leave it to others. It is also time to build the institutions for the globalized information society to come, in a world-wide collaboration, instead of starting a global war of information systems.

Appendix: Why Mass Surveillance Does Not Work

These days, it is often claimed that we need massive surveillance to ensure a high level of security. While the idea sounds plausible, I will explain, why this approach cannot work well, even when secret services have the very best intentions, and their sensitive knowledge would not be misused. This is a matter of statistics—no method is perfect.

For the sake of illustration, let us assume there are 2000 terrorists in a country with 200 million inhabitants. Moreover, let us assume that the secret service manages to identify terrorists with an amazing 99 % accuracy. Then, there are 1 % false negatives (type II error), which means that 20 terrorists are not detected, while 1980 will be caught. The actual numbers are much smaller. It has been declared that 50 terror acts were prevented in about 12 years, while a few terrorist attacks could not be stopped (although the terrorists were often listed as suspects).

It is also important to ask, how many false positives ("false alarms") do we have? If the type I error is just 1 out of 10,000, there will be 20,000 wrong suspects, if it is 1 permille, there will be 200,000 wrong suspects, and if it is 1 %, it will be 2 million false suspects. Recent figures I have heard of on TV spoke of 8 million suspects in the US in 1996, which would mean about a 4 % error rate. If these figures are correct, this would mean that for every terrorist, 4000 times as many innocent citizens would be wrongly categorized as (potential) terrorists.

Hence, large-scale surveillance is not an effective means of fighting terrorism. It rather tends to restrict the freedom rights of millions of innocent citizens. It is not reasonable to apply surveillance to the entire population, for the same reasons that it is not sensible to perform a certain medical test on everybody. There would be millions of false positives, i.e. millions of people who would be wrongly treated, with negative side effects on their health. For this reason, patients are tested for diseases only if they show symptoms of concern.

In the very same way, it creates more harm than benefit if every person is screened for terrorism. This causes unjustified discrimination and harmful self-censorship at times, where innovative and unconventional ideas are needed more than ever. It will impair the ability of our society to innovate and adapt, thereby promoting instability. Thus, it is time to pursue a different approach, namely to identify the social, economic and political factors that promote crime and terrorism, and to change these factors. Just two decades back, we saw comparatively little security problems in most modern societies. Overall, people tolerated each other and coexisted peacefully, without massive surveillance and policing. We were living in a free and happy world, where people of different cultural backgrounds respected each other and did not have to live in fear. Can we have this time back, please?

Further Reading

1. Type I and type II errors, see https://en.wikipedia.org/wiki/Type_I_and_type_II_errors

2. Qualified Trust, not Surveillance, is the Basis for a Stable Society, http://scitation.aip.org/content/aip/magazine/physicstoday/news/10.1063/PT.4.2508 and http://futurict.blogspot.ie/2013/06/qualifiedtrust-not-surveillance-is_6661.html

3. How to Ensure that the European Data Protection Legislation Will Protect the Citizens, http://futurict.blogspot.ie/2013/06/how-to-ensure-that-european-data.html

4. Big Data Is Opening Doors, but Maybe Too Many, http://www.nytimes.com/2013/03/24/technology/big-data-and-a-renewed-debate-over-privacy.html?pagewanted=all

5. Personal Data: The Emergence of a New Asset Class, http://www3.weforum.org/docs/WEF_ITTC_PersonalData-NewAsset_Report_2011.pdf

6. The Global Information Technology Report 2008–2009 Mobility in a Networked World, http://www.weforum.org/pdf/gitr/2009/gitr09fullreport.pdf

7. Statement by Vice President Neelie Kroes "on the consequences of living in an age of total information" 04/07/2013, http://europa.eu/rapid/press-release_MEMO-13-654_en.htm?locale=en

8. US Consumer Data Privacy Bill of Rights, http://www.whitehouse.gov/sites/default/files/privacy-final.pdf

9. "Daten sind die Goldminen der Zukunft", http://www.tagesanzeiger.ch/digital/internet/Daten-sind-die-Goldminen-der-Zukunft/story/15123963

10. In Angst um die Ordnung, http://www.tagesanzeiger.ch/ipad/zuerich/In-Angst-um-die-Ordnung/story/31361618

11. Post Privacy: "Der Geist ist aus der Flasche", http://www.taz.de/!131892/

10

Big Data Society: Age of Reputation or Age of Discrimination?

This chapter was first published in the FuturICT blog on September 22, 2014, see http://futurict.blogspot.ch/2014/09/big-data-society-age-of-reputation-or.html, and is reproduced here with minor stylistic improvements.

If we want Big Data to create societal progress, more transparency and participatory opportunities are needed to avoid discrimination and ensure that they are used in a scientifically sound, trustable, and socially beneficial way.

Have you ever "enjoyed" an extra screening at the airport because you happened to sit next to someone from a foreign country? Have you been surprised by a phone call offering a special service or product, because you visited a certain webpage? Or do you feel your browser reads your mind? Then, welcome to the world of Big Data, which mines the tons of digital traces of our daily activities such as web searches, credit card transactions, GPS mobility data, phone calls, text messages, Facebook profiles, cloud storage, and more. But are you sure you are getting the best possible product, service, insurance or credit contract? I am not.

Like every technology, Big Data has some side effects. Even if you are not concerned about losing your privacy, you should be worried about one thing: discrimination. A typical application of Big Data is to distinguish different kinds of people: terrorists from normal people, good from bad insurance risks, honest tax payers from those who don't declare all income .... You may ask, isn't that a good thing? Maybe on average it is, but what if you are wrongly classified? Have you checked the information collected by the Internet about your name or gone through the list of pictures *Google* stores about you? Even more scary than how much is known about you is the fact that there is quite some information in between which does not fit. So, what if you are stopped by border control, just because you have a similar name as a criminal suspect? If so, you might have been traumatized for quite some time.

Where does the problem originate? Normally, the groups of people to distinguish are overlapping—their data points are not well separated. Therefore, mining Big Data comes with the statistical problem of false positives and false negatives [1]. That is, some people get an unintended advantage, while others suffer an unfair disadvantage—an injustice hard to accept. Even with the overly optimistic assumption that the data mining algorithm has an accuracy of 99.9 %—when applied to 200 million people, there are hundreds of thousands of people who will experience a wrong treatment. In medicine, the approach of mass screenings is therefore highly controversial [2]. Are you willing to sacrifice your breast or prostata for a wrongly diagnosed cancer? Probably not, but it happens more often than you think.

Similarly, tens of thousands of honest people are unintentionally mixed up with terrorists. So, how can you be sure you are getting your loan for fair conditions, and do not have to pay a higher interest rate, just because someone in your neighborhood defaulted? Can you still afford to live in an interesting multiethnic neighborhood, or do you have to move to another neighborhood to get a reasonable loan? And what about the tariff of your health insurance? Will you have to pay more, just because your neighbors do not go jogging? Will we have to put pressure on our Facebook friends, colleagues, and neighbors, just to avoid possible future discrimination? And what would be the features that play out positively or negatively? How much sweet lemonade on our credit card bill will be acceptable to our health insurance? Is it ok to drink a glass of wine, or better not? What about another cup of coffee or tea? Can we still eat meat, or will we get punished for it with higher monthly rates? Would there be a right way of living at all, or would just everyone be discriminated for some behavior, while perhaps getting rewards for others? The latter is surely the case.

This might be fine, if everybody were to benefit in one way or another, but unfortunately this is rather unlikely. Some would be lucky and others would be unlucky, i.e. inequality would grow. But similar to stock markets, it would be difficult to tell before-hand, who would benefit and who would lose out. This is so, not just because of the random distribution of individual properties, but also because the parameters of the data mining algorithms can be determined only with a limited accuracy. However, even tiny parameter changes may produce dramatically different results (a fact known as "sensitivity" or "butterfly effect") [3]. In other words, while the miners of Big Data may pretend to take more scientific, better and fairer decisions, the results will often have a considerable amount of arbitrariness. Many data miners probably don't know about this or don't care. But the fact that lots of al-gorithms produce outputs without warnings of their limitations creates a dangerous overconfidence in their results. Moreover, note that the choice of the model can be even more critical than the choice of parameters [4]. That's basically why people say: "Don't believe a statistics that you haven't produced yourself."

The problem is reminiscent of the experiences made with financial innovations. People used models without thoroughly questioning their validity. It was discovered too late that financial innovations may have negative effects and destabilize the markets. One example is the excessive use of credit default swaps, which package risks in ways that buyers don't seem to understand any-more. The consequence of this was a financial meltdown that the public has to pay for at least for another decade or two. It is no wonder that trust in the financial system dropped dramatically, with serious economic implications (no trust means no lending). This time, we should not make the same mistakes, but rather use Big Data in a trustworthy, transparent, and beneficial way. To reap the benefits of personalized medicine, for example, we need to make sure that personal medical data will not be used to the disadvantage of patients who are willing to share their data in favor of creating a public good—a better understanding of diseases and how to cure them.

In fact, we have worked hard to overcome discrimination of people for gender, race, religion, or sexual orientation. Should we now extend discrimination to hundreds or even thousands of variables, just because Big Data allows us to do so? Probably not! But how can we protect ourselves from such discrimination? In order to avoid that the information age becomes an age of discrimination fueled by Big Data, we need *informational justice*. This includes to establish (1) suitable quality standards like for medical drugs, (2) proper testing, and (3) fair compensation schemes. Otherwise people will quickly lose trust in Big Data. This requires us to decide what collateral damage for individuals would be considered tolerable or not. Moreover, we need to distinguish between "healthy" and "unhealthy" innovations, where "healthy" means innovations that produce long-term benefits for the economy and society (see Information Box). That is, the overall benefit should be bigger than the disadvantage caused by false positives, such that the corresponding individuals can be compensated for unfair treatments.

There are two fundamentally different ways to ensure a "healthy" use of Big Data and allow victims of discrimination to defend their interests. The classical approach would be to create a dedicated government agency or institution that establishes detailed regulations, in particular quality standards, certification procedures, and effective punitive schemes for violations. But there is a second approach—one that I believe could be more effective for companies and citizens than complicated legal and executive procedures. This framework would be based on next-generation reputation systems creating feedback loops that support self-regulation.

How would such a next-generation reputation system work? The proposal is to establish a *Global Participatory Plattform* [5], i.e. a public store for models and data. It would work a bit like an *appstore*, but people and companies could upload not only *apps*. They could also upload data sets, algorithms (e.g. statistical methods, simulation models, or visualization tools), and ratings. Everybody could use these data sets for free or for a fee, and annotate user feedbacks. It would be as if we were able to submit not only queries to *Google*, but also algorithms to determine the answers. In this way, we could better control the quality of results extracted from the data.

So, assume we would store all data collected about individuals in a data bank (for reasons of data security, a decentralized and encrypted storage would be preferable). Moreover, assume that everyone could submit algorithms to be run on these datasets. The algorithms would be able to perform certain operations within the bounds of privacy laws and other regulations. For example, they could generate aggregate information and statistics, while privacy-invasive queries violating user consent would not be executed. Moreover, if executable files of the algorithms used by insurance or other companies using Big Data were uploaded as well, it would allow scientists and citizens to judge their statistical properties and verify that undesirable discrimination effects are below commonly accepted thresholds. This would ensure that quality standards would be met and continuously improved.

The advantages of such a transparent and participatory approach are multifold for business, science, and society alike: (1) results can be verified or falsified, thereby uncovering possible methodological issues, (2) the quality of Big Data algorithms and data will increase more quickly, (3) "healthy" innovation and economic profits will be stimulated, (4) the level of trust in the algorithms, data and conclusions will increase, and (5) an "information ecosystem" will be grown, creating an enormous amount of new business opportunities, to fully unleash the potential of Big Data.

I fully agree with the US Consumer Data Privacy Bill of Rights [6] stating that *"trust is essential to maintaining the social and economic benefits that networked technologies bring to the United States and the rest of the world."* A report on personal data as a new asset class, published by the World Economic Forum, therefore suggests a "New Deal on Data" [7]. This includes establishing a data ecosystem that creates a balance between the interest of companies, citizens, and the state. Important elements of this would be: transparency, more control by citizens over their personal data, and the ability for individuals to participate in the value generated with their personal data.

This has implications for the design of the *Global Participatory Platform* I am proposing. Data collected about individuals would be stored in a personal data purse. Individuals could add and comment the data, have them corrected, if factually wrong, and determine, who could use them for what kind of purpose, to meet the regulations regarding privacy and self-determination. When personal data are used, both the user and the company that collected the data would earn a small amount, triggering micropayments. Finally, to keep misuse of data and malicious applications on a low level, there would be a certain reputation system, which would act like a *social immune system*.

Reputation and recommender systems are quickly spreading all over the Web. People can rate products, news, and comments. In exchange, *Amazon*, *eBay*, *TripAdvisor* and many other platforms offer recommendations. Such recommendations are beneficial not only for the user, who tends to get a better service, but also for a company offering the product or service, as higher reputation allows it to take a higher price [8]. However, it is not good enough to leave it to a company to decide, what recommendations we get, because then we don't know how much we are being manipulated. We want to look at the world from our own perspective, based on our own values and quality criteria. It would be terrible if everyone ended up reading the same books and listening to the same music. Therefore, it is important that recommender systems do not undermine socio-diversity.

Diversity is an important factor for innovation, social well-being, and societal resilience [9]. It deserves to be protected in the very same way as biodiversity. Modern societies need a complex interaction pattern of diverse people and ideas, not average people who all do the same things. The socio-economic misery in many countries of the world is clearly correlated with the loss of socio-economic diversity. While some level of norms and standardization appears to be favorable, too much homogeneity turns out to be bad. This also implies that we need to be careful about discriminating against people who are different—such discrimination may undermine socio-diversity.

Today's personalized recommender systems endanger socio-diversity as well. They are manipulating people's opinions and decisions, thereby imposing a certain perspective and value system on them. This can seriously undermine the "wisdom of crowds" [10], which is central to the functioning of democracies. The "wisdom of crowds" requires independent information gathering and decision-making—a principle not sufficiently respected by most recommender systems [11].

How could we, therefore, build "pluralistic" reputation and recommender systems, which support socio-economic diversity, and are also less prone to manipulation attempts? First, one should distinguish three kinds of user feedbacks: facts (linked to information allowing to check them), advertisements (if there is a personal benefit for posting them), and opinions (all other feedbacks). Second, user feedbacks could be made in an anonymous, pseudonymous, or personally identifiable way. Third, users should be able to choose among many different reputation filters and recommender algorithms. Just imagine, we could set up the filters ourselves, share them with our friends and colleagues, modify them, and rate them. For example, we could have filters recommending the latest news, the most controversial stories, the news that our friends are interested in, or a surprise filter. So, we could choose among a set of filters that we find most useful. Considering credibility and relevance, the filters would also put a stronger weight on information sources we trust (e.g. the opinions of friends or family members), and neglect information sources we do not want to rely on (e.g. anonymous ratings). For this, users would rate information sources as well, i.e. other raters. Therefore, spammers would quickly lose reputation and, with this, their influence on recommendations made.

In sum, the system of personal reputation filters would establish an "formation ecosystem," in which increasingly good filters will evolve by modification and selection, thereby steadily enhancing our ability to find meaningful information. Then, the pluralistic reputation values of companies and their products (e.g. insurance contracts or loan schemes) would give a pretty differentiated picture, which can also help the companies to develop better customized and more successful products.

In conclusion, I believe it's high time to create suitable institutions for the emerging Big Data Society of the twenty-first century. In the past, societies have created institutions such as public roads, parks, museums, libraries, schools, universities, and more. But information is a special resource: it does not become less, when shared, and it can be shared as often as we like. In fact, our culture results from what we share. At the moment, however, the world of data is highly proprietary and fragmented. It's as if every individual owned a few words but had to pay for using all the others, and some words could not be used at all for proprietary reasons. Obviously, such a situation is not efficient and does not make sense in an age where data are increasingly important.

Business and politics have pushed hard to remove barriers to the free trade of goods—it is now time to remove the obstacles to the global use of data. Providing access to Big Data would unleash the power of information for business, politics, science and citizens. Access to Big Data is surely needed for science to provide a good service to society [12, 13]. In the past, reading and writing was a privilege, which came with personal advantages. But public schools opened literacy to everyone, thereby boosting the development of modern service societies. In the very same way we could now boost the emerging digital society by promoting digital literacy and investing into transparent, secure, participatory and trustworthy information and communication systems [14]. The benefits for our societies would be huge!

10.1 Information Box: How to Define Quality Standards for Data Mining

Assume that the individuals in a population of N people fall into one of two classes. Let us consider people of kind 1 "desirable," e.g. honest citizens, good insurance risks) and people of kind 2 "undesirable" (criminals, bad insurance risks, etc.). We represent the number of people *classified* as kind 1 and 2 by N_1 and N_2, respectively. Let the rate of false positives, that is individuals who are faced with unjustified discrimination, be given by α, and the rate of false negatives be β. Then, the *actual* number of people of kind 1 is $(1 - \beta)N_1 + \alpha N_2$, and the *actual* number of people of kind 2 is $(1 - \alpha)N_2 + \beta N_1$. Furthermore, assume that the classification is creating an advantage of $A > 0$ for people classified as kind 1, but a disadvantage of $-D < 0$ for people classified as kind 2. Then, each false positive classified person has a double disadvantage of $-(A + D)$, because he or she should have received the advantage A while suffering the disadvantage $-D$. This will be considered unfair and question the legitimacy of the procedure. False negatives, in contrast, those who are classified "desired" but are in fact "undesired", enjoy a double advantage of $(A + D)$. They may also create an extra damage $-E$ to society. Overall, the classification produces a gain of $G = N_1[(1 - \beta)A + \beta(A + D)]$ to individuals classified to be of kind 1 and a cost of $C = -N_2[(1 - \alpha)D + \alpha(D + A)]$ to individuals classified to be of kind 2. The overall benefit to society would be $B = G - C - E$. Unfortunately, there is no guarantee that it would be positive.

To demonstrate this, let us assume a business application of Big Data, in which the economic profit P (e.g. by selling cheaper insurance contracts to people of kind 1) is a fraction f of the gain, i.e. $P = fG$. If applied to many people, the application may be profitable even if the fraction $f < 1$ is quite small. Moreover, from the point of view of a company, discrimination may be rewarding even if it has an overall disadvantage to people (i.e. if the overall benefit B is negative). This is because a company typically cares about its own profits and its customers, but not everybody else. Clearly, if some insurance contracts get cheaper, others will have to be more expensive. In the end, people with high risks will not be offered insurances anymore, or only at an unaffordable price, so some victims of accidents may not be compensated at all for their damage.

Even if B is positive, the profit P may be smaller than the unjust disadvantage U, which is the price that false positives have to pay. Such a business model would create a situation that I will call a *"discrimination tragedy,"* where citizens have to pay the price for economic profits, even though they are not getting a good service in exchange.

It is, therefore, in the public interest to establish binding standards for the "healthy" use of Big Data algorithms, regulating the required predictive power and the acceptable values of α, D, B and

U. A cost-benefit analysis suggests to demand $B > 0$ (there is a benefit) and $B > U$ (the benefit is high enough to compensate for unjust treatments). Moreover, αN_1 and D should be below some acceptable thresholds. Today, these values are often unknown, and that means we have no idea what economic and societal benefits or damages are actually created by current applications of Big Data.

References

1. C. Chivers, How likely is the NSA PRISM program to catch a terrorist?, bayesianbiologist, http://bayesianbiologist.com/2013/06/06/how-likely-is-the-nsa-prism-program-to-catch-a-terrorist/ and http://futurict.blogspot.ch/2013/06/why-mass-surveillance-does-not-work.html

2. G. Gigerenzer, W. Gaissmaier, E. Kurz-Milcke, L.M. Schwartz, S. Woloshin, Helping doctors and patients make sense of health statistics. Psychol. Sci. Public Interest **8**(2), 53–96 (2008)

3. I. Kondor, S. Pafka, G. Nagy, Noise sensitivity of portfolio selection under various risk measures. J. Bank. Finance **31**(5), 1545–1573 (2007)

4. T. Siegfried, Odds are, it's wrong. Sci. News. **177**(7), 26 (2010), http://www.sciencenews.org/view/feature/id/57091/description/Odds_Are_Its_Wrong. (J.P.A. Ioannidis (2005) Why most published research findings are false, *PLoS Medicine* **2**(8): e124)

5. S. Buckingham Shum, K. Aberer, A. Schmidt, S. Bishop, P. Lukowicz et al., Towards a global participatory platform (2012) Democratising open data, complexity science and collective intelligence. EPJ Spec. Top. **214**, 109–152 (2012)

6. The White House, Consumer data privacy in a networked world: A framework for protecting privacy and promoting innovation in the global digital economy (2012), http://www.whitehouse.gov/sites/default/files/privacy-final.pdf

7. World Economic Forum, Personal data: The emergence of a new asset class (2011), http://www3.weforum.org/docs/WEF_ITTC_PersonalDataNewAsset_Report_2011.pdf

8. W. Przepiorka, Buyers pay for and sellers invest in a good reputation: More evidence from eBay. J. Soc.-Econ. **42**, 31–42 (2013)

9. S.E. Page, *The Difference* (Princeton University Press, Princeton, 2007)

10. J. Lorenz, H. Rauhut, F. Schweitzer, D. Helbing, How social influence can undermine the wisdom of crowd effect. *Proceedings of the National Academy of Sciences of the USA* **108**(28), 9020–9025 (2011)

11. T. Zhou, Z. Kuscsik, J.-G. Liu, M. Medo, J.R. Wakeling, Y.-C. Zhang, Solving the apparent diversity-accuracy dilemma of recommender systems. Proc. Natl. Acad. Sci. U S A **107**, 4511–4515 (2010)

12. B.A. Huberman, Big data deserve a bigger audience, Nature **482**, 308 (2012)

13. F. Berman, V. Cerf, Who will pay for public access of research data? Science **341**, 616–617 (2013)

14. D. Helbing, Economics 2.0: The natural step towards a self-regulating, participatory market society. Evolut. Inst. Econ. Rev. **10**(1), 3–41 (2013)

11

Big Data, Privacy, and Trusted Web: What Needs to Be Done

*This is second part of the paper: D. Helbing and **S. Balietti**, From social data mining to forecasting socio-economic crises, EPJ Special Topics **195**, 3--68 (2011), see http://link.springer.com/content/pdf/ 10.1140%2Fepjst%2Fe2011-01401-8.pdf © EDP Sciences, Springer-Verlag 2011. It is reprinted here with kind permission and with minor stylistic improvements.*

This perspective paper discusses challenges and risks of the information age, and the implications for the information and communication technologies that need to be built and operated. It addresses ethical and policy issues related with Big Data and how procedures for privacy-preserving data analyses can be established. It further proposes a concept for a future, self-organizing and trusted Web and discusses recommended legal regulations as well as the infrastructure and institutions needed.

11.1 Ethical and Policy Issues Related with Socio-Economic Data Mining

Large-scale data mining is opening up previously unimaginable, new perspectives for science and, of course, even more for business. At the same time, it affects fundamental rights of individuals in ways, which are hard to fully oversee. Among these, the right of privacy is surely one of the most endangered, but it is not the only one. Such risks result not only from single research or data-mining activities. They arise in particular from the combination of singular observations in larger datasets, which contain more and more information, and are capable eventually to depict accurate personal profiles. With these giant data conglomerates at one's disposal, making sense of unpersonalized and apparently irrelevant information is easier than one could think [1]. However, it is still not clear what the implications of developing such *informational cornucopias* are. In the meantime, megadata centers run by private companies and national security agencies are spreading [2, 3, 4, 5, 6]. Intel, the largest CPU manufacturer in the world, has declared that already by 2012 megadata centers will account for 20–25 % of its server chip sales [7].

In the following sections, we will discuss ethical aspects of building gigantic supercomputing ICT facilities for large-scale data mining, as the ones mentioned before. Our analysis will be primarily based on and guided by a literature review of ethical research in the social sciences. The approach followed can be characterized essentially (but not exclusively) by a positivist and structuralist standpoint, and our discussion will concentrate mainly on privacy issues. However, in Sect. 11.1.5 we will consider other ethical concerns inherently related to large-scale data-mining activities. Further ethical issues related to social supercomputing are addressed in Ref. [8].

11.1.1   A Source-Based Taxonomy of Available Personal Information

Given that today, more information is available about us than we are usually aware of, let us start the discussion of ethical issues with a picture of the personal data traces almost everyone leaves most of the time. The following paragraphs provide a non-exhaustive taxonomy of available data organized by data sources.

Data in Public Registries Data belonging to this category is generally already available to the public, or available after paying a small fee to public institutions.

- phone books,
- land registries,
- car plate registries,
- health data,
- salary registries (available primarily for the public sector)
- tax data (public in the US),
- religious confession,
- social security and passport numbers.

Data Generated by Electronic Services Today, the correct and efficient functioning of our everyday lives is more or less dependent on a few essential services, which are increasingly supported by ICT and electronic infrastructures. This means that, by using such services, a lot of data are automatically generated as by-product. Data in this category are usually available only to certain public institutions and/or some private companies providing these services.

- phone call logs,
- flight passenger information (such as e-mail addresses, credit cards, etc., particularly for flights to the USA),
- bank account data,
- credit card numbers,
- money transactions (e.g. Swift system),
- consumer data ("people who bought X have also bought Y"),
- behavioral analyses.

Data Generated by Internet Activities "Look but do not touch" was considered a wise advice to follow when entering unknown environments. However, in the Internet, this is no longer sufficient. The sheer surfing activity, without any content and without accessing any service requiring authentication, e.g. reading certain news, is enough to generate a wide range of differentiated digital traces. These traces are stored on private remote servers as well as on the local drives. This includes

- Internet service provider logs (e.g. IP and MAC addresses),
- logs of remote access to phones and computers,
- browser history,
- browser cache,
- cookies,
- search queries, and
- click streams.

Data from Portable Devices In many social strata, the everyday usage of portable devices is becoming a wide-spread habit. The current integration trend makes portable devices more and more interconnected with each other through wireless communication networks. This facilitates the spatial tracking of persons via location data, which are exchanged by their devices. Such data include

- GSM, UMTS, and GPS location data,
- WLAN/WiFi open hot spots,
- bluetooth devices,
- RFID data,
- car transponders for automated highway toll payment systems,
- electronic badges (e.g. for conferences [9], hotel rooms, etc.)

Moreover, the large availability of peer-to-peer connections and Internet access points increases the risk of security breaches and data leaks, especially when these devices are used by people unaware of their vulnerabilities.

Finally, the portability of such media introduces the risk of loss of the device itself and consequently of all data stored in them. Given the ongoing miniaturization process and the steady improvements in capacity, the privacy concerns arising from the lack of encryption or other data-protection techniques for such devices are real. This concerns, in particular,

- cameras and video recorders,
- cell phones,
- electronic organizers and smart phones,
- laptops,
- flash memory cards and external hard drives, and
- smart multimedia players.

Unauthorized Content Captured from Diverse Multimedia Devices Individual actions that reveal people's lifestyles may be recorded in both public and private venues and made public at any time and without any previous warning. This concern is becoming increasingly more real due to the integration of multimedia contents into global projects such as Street View, and the success of photo and video online repositories. This concerns

- uploaded content on social websites (e.g. embarrassing party snapshots or videos),
- Google Street View photographs,
- public webcams.

User-Generated Contents Many users "voluntarily" share personal opinions or even detailed personal information on their online profiles. Whether they are aware of all the risks of this practise is not entirely clear, but the material is sufficient to identify political, religious and/or sexual preferences of many Internet users. This concerns

- blogosphere data (forums, blogs, chats, etc.),
- the archive of mailing lists or discussion groups,
- keyword scans of free mail accounts,
- social network data.

Security Data   Under the key issue of security, people were willing or forced to reduce the range of their personal freedoms, with consequences often also for personal privacy. This can happen through an explicit disclosure of personal data, e.g. filling in a security form to entering a foreign country or through accessing a given service, or tacitly, e.g. through public surveillance cameras.

- video surveillance (CCTV),
- face recognition data,
- biometric data,
- audio recordings, directional microphone recordings,
- phone call surveillance,
- speed radar photographs,
- scanned items and body scans at airports,
- security forms that must be filled in.

Intercepted Data   From very basic to very sophisticated techniques, despite this may be for illicit purposes, electronic communications can be intercepted. Examples include

- network eavesdropping (emails traffic, phone calls, etc.),
- identity theft,
- hardware trojans,
- software trojans,
- the physical analysis of variations in electromagnetic fields of wireless devices (keyboard and mouse) and of computer screens,
- the monitoring of fluctuations in the electricity consumption of electronic devices.

While the above lists are probably not complete, it is obvious that the combination of only some of the above data can eliminate privacy to a large extent. Modern information services give a striking picture of this (see e.g. [10]). On the one hand, they show how much information can easily be gained about a single person (contact data, pictures, videos, news, etc.). On the other hand, they illustrate how easily wrong information not related to the person searched for is mixed between correctly retrieved information. Therefore, we will discuss below whether privacy is just an outdated concept, or whether it is crucial for the functioning of democratic societies.

11.1.2 Why Would the Honest be Interested to Hide?

When it comes to private data, some people suggest that privacy is mainly in the interest of dishonest, criminal, or perverted people. In the following, we will argue that this is a dangerous misconception. Privacy has been granted not as a concession of the state to the individual, but because a modern society needs it in order to flourish.

Although different in several respects, commercial confidentiality may serve as a useful illustration to explain why privacy is an essential requirement for people. For example, if confidentiality is eliminated, there would be no incentive for companies to invest into expensive long-term innovations, which pay off only through a winning margin. It would be so much more economic to copy inventions of others as soon as they occur. (There would not be such a fierce discussion about copyright protection/patent enforcement, if this would not be the case.) Secrecy and confidentiality are needed to gain a competitive advantage (in particular in time) that makes innovation commercially profitable. There are two other interesting points about innovation:

- Innovation usually starts off in a minority position [11]. In the beginning, there are a few supporters and customers only. In other words, there is little innovation without the existence of minorities.
- As is known from evolutionary theory, innovation thrives best when there is a large diversity of variants [12]. In other words, diversity or "pluralism" is the motor driving innovation. If we

just followed the majority or aimed at being "normal" (the average), the innovation rate and, with this, our adaptability to changing (environmental) conditions would suffer. This is actually the reason why totalitarian regimes are sooner or later destined to fail.

These principles can also be transferred to individuals. Without privacy, pluralism is in danger, as the following lab experiment shows [13]: Experimental subjects had to guess the correct answer to a factual question such as "How many murderers occurred in the year 2006?" They received a certain amount of money, whenever their answer was close enough to the correct one. In one setup, they decided several times without any information feedback; in another setting, they were informed about the estimates of the others. In the first round, the variation of answers was high, but the correct answer was always within the range of answers and was usually well approximated by the geometric mean value of all estimates. When information feedback was provided, the answers converged over time, which may be an indication that the right answer was identified. Instead, however, it often happened that the relevant spectrum of answers did not contain the correct answer. In other words, social imitation created herding effects, which were often misleading.[1]

The financial crisis is probably an example for such herding effects, which led to extremely expensive mistakes. Herding-related mistakes would become even more likely, when people were put under pressure to conform with frequent opinions or behaviors, and as the above experiment shows, even when they would only be exposed more often to other opinions than they used to be. This

---

[1] Note that taking the wrong decision occurred here even without social pressure, while it is known since the famous Asch experiment that individuals give predominantly wrong answers (against their own judgement), if the people before them do so [14].

applies as well to many current Web services and recommender systems, which reinforce dominating opinions.

Revealing private data would increase this tendency of conformism enormously and would have other unwanted side effects, as the following points indicate:

1. If "normal" behavior was considered the ideal and individualism was to be discouraged, life would become more predictable, but for sure also much more boring.

2. Conformism implies a danger of discrimination (for having a certain religious belief, age, gender, disease, sexual preference, etc.; it is not without reason that Americans apply for jobs without a birth date and without a photograph). It is well-known that minorities need protection. One must be aware that it is usually minorities who create the concepts and lifestyles of tomorrow, and that it is hard to say in advance, which ones it will be. The minority behavior that eventually becomes majority largely depends on environmental changes and historical developments. A society, therefore, needs to have a pool of minorities to successfully adapt to the challenges and opportunities of the future. Minorities are an indispensable ingredient in the process for evolutionary innovation [15, 12].

3. The majority behavior of today may be a minority behavior of tomorrow. What is normal today may be perceived as abnormal tomorrow. For example, it is hard to predict how we will think in the future about the appropriateness of certain kinds of food we eat or the environmental and labor conditions under which purchasable goods are produced. Hence, nobody can be certain that his or her current behavior would be considered acceptable or within the norm in the future. Social norms are continuously changing [16]. For example, in the 1960s, the values of society were changing dramatically, and the establishment got under enormous pressure.

There are many other examples, such as racial segregation, which was considered "normal" by many people in the past, but is seen totally different nowadays.

4. Private data could be misused by companies. For example, insurance companies have an interest to offer cheap contracts to the majority of people and to charge minorities for special risks (e.g. inborn or past health risks, or higher hospital costs of women giving birth to children). This, however, clearly undermines solidarity.

5. Publicly available data could also be misused towards criminal means. For example, the city of Oakland releases information on where and when arrests are made, which is later on displayed on a private website [17]. From that website, it was possible for criminals to infer the police's tactics, patrolling times and other valuable information [18].

6. Companies start charging money to people who want certain private information to be deleted [23, 19, 20, 21, 22]. A recent newspaper article even predicts that privacy in the future will be a privilege of the rich [24].

7. Disclosing the wealth of people explicitly or implicitly (e.g. through Street View services) can endanger individuals and increases the chance that they may become victims of crime. Therefore, being rich may become less rewarding, and all the private initiative, innovation and commitment leading to it as well.

8. Generally, people with professions that require them to take unpopular decisions sometimes (such as judges, policemen, or teachers) sometimes need a certain degree of protection of their private sphere. Otherwise, they will not be able to perform their job effectively and end up doing what pleases those they are supposed to judge.

9. People may no longer be able to seek forgiveness for mistakes they have made, if information about them remains publicly

accessible forever [25]. In the past, after a reasonable punishment, depending on the gravity of the misconduct, the policy of societies was to forget about them. In the worst case, delinquents would still be able to emigrate to other countries, where nobody knew them, paying with abandoning their hometown, a high price for getting a *second chance*. Now, however, wherever one may go, the digital traces left behind will follow him or her. This is not necessarily bad, but it certainly requires a savvier society that is able to remember and forgive at the same time. As Thomas Szasz said "The stupid neither forgive nor forget; the naive forgive and forget; the wise forgive but do not forget." Without an adequate mix of tolerance and solidarity, the ability of a society to (re-)integrate people could be seriously undermined. Outcasts would only have a chance to find friends among other outcasts. As a consequence, this would fragment society into a variety of subsocieties—a trend, which is observed already.

10. Whenever a huge amount of personal information is available, individuals, private businesses or public institutions may try to infer individuals' behaviors, preferences and attitudes and to classify them according to certain profiles. This tendency is as strong as dangerous, since there is no such thing as an accurate classification. Moreover, in the presence of information asymmetries, which are extremely common in everyday life (such as market exchange, buyer/seller interactions, insurance contracts, bank operations, job interviews, etc.), an inappropriate or wrong classification may be hard to correct or oppose to. Moreover, it may affect the lives of people in manifold and unexpected ways, given the high degree of interconnectedness of different services. In the worst case, it can even lead to *circuli vitiosi* from which people cannot escape. For example, missing an installment for leasing a car once could mark somebody as bankrupt.

This label would prevent this person from getting future loans, which he or she would need in case of temporarily financial recovery. However, it could lead to even more absurd situations. For example, by skipping one installment, the system would automatically register the fact "interruption of contract", and tag one's profile with a negative label. Ironically, the real motivation behind the fact "interruption of contract" could even be that the entire amount of money due was paid at once, without waiting until the contract expired.

The above example is real, and wrong classifications like these are already happening. But that is not yet the worst possible scenario one can imagine. In fact, we must be aware that any form of classification introduces elements for discrimination, because the "labels" are often not adequate and not mutually agreed on [26]. Classifications (whether justified or not) create peer groups and may seriously undermine the basis of cooperation and shared norms in our society. They may also cause unnecessary conflicts [27].

11.   As it becomes possible to learn quickly what kind of people we are interacting with and what they do and think, this will undermine an independent judgement of their qualities (and weaknesses). Rather than giving everybody a fair chance to find the right kind of friends, people might end up stigmatized and socially excluded. People need to be protected from intolerance, mobbing, blackmailing and bribery. To live in peace, people often choose to segregate from others. Given the availability of a lot of personal information to everyone, however, the Internet does not allow this anymore. In this respect, it is important to note that undermining the mechanism of voluntary seclusion can seriously affect the cooperation among people, to the disadvantage of everybody [28, 29].

12. The more the Internet knows about everyone, the closer we get to a situation where we can effectively read other people's minds. Such a situation, however, would potentially generate a lot more conflicts than we have today.

13. It must also be noted that having more information freely available does not necessarily lead to a more transparent, fairer or better society. In an information-rich environment, people spend only a short time on a certain subject, and it easily can happen that people get a wrong impression based on such a *pars pro toto* approach (assuming that the first or a randomly picked piece of information would be representative for the full information). Therefore, large amounts of information can promote misjudgements of a person's behavior by the media and by public opinion [30]. Such reputational effects are difficult to correct, particularly as rectifications (e.g. when a suspect in a crime case has been found innocent) are often poorly noticed. This may have a serious impacts on individual lives.

14. When all people have access to the same information at the same time (and at negligible costs), this may lead to negative feedback effects such as herding effects. A typical example is the information about a traffic jam, which is provided to everybody via the public news. One can easily imagine that this leads to over-reactions of drivers to the news and, thereby, to overloaded alternative roads, while the originally congested route may become underutilized. A possible solution of this problem is to provide user-specific information according to probabilistic rules [31] or to overlay randomness to the information signal [32].

15. Systems where a high degree of transparency has already been implemented for years have shown to become more sensitive to sudden regime shifts. Examples are market hysteria and volatility clustering phenomena, which can cause failure avalanches. In some cases, transparency on the producer

side can also facilitate the establishment of tacit collusive practises, as it has been found in online markets, auctions, and laboratory experiments [33, 34, 35].

16. Decisions to reveal private information may even spread in an "epidemic" way. For example, if someone decides to provide access to personal data (such as GPS car tracking data, in order to get a cheaper car insurance contract), this can deteriorate the conditions and potentially restrict the options for those, who do not want to give up their privacy. In other words, revealing one's own data can have an impact on other people who chose not to do so, but who are eventually forced to provide private information in order to maintain the same contract conditions and the same price they had to pay before. This also applies to private health insurances, for example.

17. The data on the servers of certain Internet companies probably include more details about us than what our friends and partners, and maybe even we ourselves know about us. However, when knowing the preferences of customers, companies may try to manipulate their choices, and possibilities to do so may increase with personalized recommendations (special offers may even have addictive effects). As it becomes possible to shape the customers expectations, this is likely to decrease the willingness of producers to tailor products and services to the needs that customers really have. In fact, due to the "economies of scale", businesses have a natural interest in providing a number of standardized products.

18. Finally, recent scientific studies indicate that pluralism in a society may get lost, as new technologies change the parameters of the opinion formation dynamics [36]. Socio-diversity and its benefits (as outlined above), may easily get lost in favor of conformism and monoculture. It requires the mechanism of individualization, i.e. the desire to be different from others. Therefore, technologies or circumstances promoting

conformism may seriously endanger the basis of democracies. In fact, the danger to suppress minority opinions and preferences increases as large datasets containing private information are centrally stored, and as it becomes possible to connect different kinds of datasets. It is clear that knowledge implies power, and it would be naïve to think that people would not use it. In fact, there are many examples of misuse of private data (see the section on cyber-risks below). It would be surprising if organized crime did not try to get access to Google's data. One of the few laws of social systems, which have been confirmed again and again is: "Anything that can go wrong ... generally does go wrong sooner or later," This raises concerns, as today's information systems probably *would* give anyone the power to damage today's pluralistic societies, if whoever really wanted. After all, the Internet contains more sensitive information and about a larger number of people than secret services of totalitarian states ever had. In addition, experience tells us that no database is absolutely safe. In 2009, for example, several large sensitive datasets were stolen from public institutions in Great Britain, where they should have been well protected [37].

Therefore, the storage and processing of large datasets of socioeconomic activities is a very sensitive issue. They certainly have the potential to harm pluralistic societies. The interests of individuals (such as privacy) and companies (such as details of their business) *must* be protected. Therefore, it is necessary to address cyber-risks and ethical issues by scientific, legal and technological means. The following sections provide guidelines on how this could be done.

11.1.3 Cyber-Risks and Trust

Big data aggregates represent much sought-after targets for cyber-criminals and pose big challenges for security experts. The Symantec Internet Security Threat Report XV [38] mentions a 100 % increase in the number of new malicious programs identified (more than 240 million in 2009) and estimates the number of Internet users (companies and individuals), who have been victims of cyber-attacks trying to steal money or confidential information, to be of the order of 360 millions. More and more attacks are aiming at *identity theft*. Sixty percent of all data breaches that revealed identities were in fact the result of hacking.

An incomplete list of the risks of using the Internet today is given below:

- data theft,
- theft of pin codes and passwords,
- identity theft,
- viruses, worms, and trojan horses (damaging software, steeling passwords, etc.),
- data manipulation,
- wrong evidence (wrong accusations),
- malicious rumors [39],
- information pollution,
- spam and unwanted advertisements.

These risks may seriously undermine the trust of people in the Internet and services based on it. For example, the theft of access data for electronic banking through phishing attacks has recently become a widespread problem. However, trust is essential for economic exchange. Systems which would not be able to effectively work without a certain level of trust include:

- electronic banking,
- e-mail,
- eBusiness,

- eGovernance, and
- social networking.

To solve the above problems, the right mixture between legal regulations and technical innovations is needed.

11.1.4 Current and Future Threats to Privacy

Whether personal data disclosure in the Internet is the result of a truly *voluntarily and deliberate* choice is rather questionable. In social research, voluntarily participation is considered a basic human right, which overlaps considerably with the principle of informed consent [40]. Moreover, European law, for example, gives people an individual right of control over personal information.

There is no unanimous definition for informed consent, but according to Diener and Crandall [41] it is "the procedure in which individuals choose whether to participate in an investigation after being informed of the facts that would be likely to influence their decision". In principle, any decision can be considered as informed consent if it has been taken after being provided with the amount of information that *a reasonable and prudent person would want to have* [42]. In the Internet this is seldom the case. In fact, it is both possible and relatively common for individuals to access Web sites without reading the terms and conditions (which may be several dozen pages long). It is also unlikely that most people would understand the full contract, while they actually have to approve this. Moreover, they are usually not given any options other than accepting the conditions in order to get the requested service or rejecting them at the cost of no service, which does not give users a reasonable choice. Under these circumstances, people may nominally give consent, but without being fully aware of or agreeing with the terms and conditions. Such a situation would not be considered to comply with informed consent [43]. This stands in contrast to a widely accepted principle in Social Science

Ethics that states that "*as far as possible, participation in sociological research should be based on the freely given informed consent of those studied*" [45, 44]. Moreover, fully informing the respondents it is not yet enough, since researchers should endeavour to make sure that the participants of an experiment have fully understood the involved risks and consequences [42]. This applies in particular for physically or mentally impaired individuals [45, 46], but cannot be ensured by the Internet [43].

Whether large data-mining companies are aware of the above mentioned ethical issues is questionable, especially when CEO's of big data-mining companies make statements about privacy such as the following one: "*If you have something that you don't want anyone to know, maybe you shouldn't be doing it in the first place*" [47]. This is worrying, because if ethical standards turned out to be insufficient at some of the fundamental places of command of the biggest data-mining companies, or if market competition would push them to pursue only the logic of profit, what would refrain them from collecting and using people's data even in illicit ways? Data-mining techniques improve every day, while regulations and control over the gathered data are lacking far behind. For example, tracking the source of collected information—once it is stored in secured and not publicly accessible databases—is virtually impossible; knowing who has access to which kind of personal data is also not possible today. Relevant to this discussion and particularly controversial is the latest case of Street View cars. For several months, these cars had been storing personal data, including passwords, credit card information and accessed email contents, which were intercepted from private WiFi networks. The incident was reported as a result of a programming error, but others have suggested that this was rather a case of WiFi sniffing, as there exists a software patent which involves intercepting data and analyzing the timing of transmission as part of the method for pinpointing user locations. At the time of writing this White Paper, the actual situation is still unclear [48, 49, 50, 51].

Also when not possessing a sophisticated and expensive data-mining system, criminals can collect illicit data easily through Web browsers, as these are daily affected by new malicious exploits (see [52]). The most common attacks are now based on a technique called "history stealing". Some websites even show this security issue to visitors [53, 54], thereby demonstrating how easy it is to extract personal surfing habits of Internet users. Scientific literature on the topic is vast, and latest studies conducted on 243,068 users found that 76 % of them were vulnerable to history detection by malicious websites. Newer browsers such as Safari and Chrome were even more affected, with 82 and 94 % of vulnerable users [55]. Unfortunately, there is yet another privacy issue related to recent generations of Web browsers: their inherently high customization capabilities have made them unique, and therefore trackable. In fact, even disabling cookies, and blocking history-stealing-like exploits, individual Web surfing can still be reconstructed by simply following the customized "fingerprint", which the browser is carrying around from site to site. This fingerprint is actually made up by all the configuration information that the browser is exposing to remote Web sites. According to the Electronic Frontier Foundation [56], information such as which plugins are installed, which fonts are available and which operating system the browser is running on, can create a unique portrait of 94 % of the visitors (for a self-demo see Ref. [57]).

Unethical or dishonest intents are not the only pitfalls glooming over online data sharing. Even in a scenario, in which one has consciously provided his or her own personal data to a company that uses them lawfully, unforeseen issues can suddenly arise. For example, such a company could be sold or merged with another one, or simply, the data could be sold, based on a change in the data-handling policies. Users are typically not notified of such changes, and they usually have no effective possibility to withdraw their data and their consent to use them. Some social network Web

pages are examples for this. In fact, because of the continuous up-
dating and modification of the terms of use [58], the Electronic
Privacy Information Center (EPIC) has filed a formal complaint to
the US Federal Trade Commission [59], and more lately U.S. Senator Charles Schumer (D-N.Y.) has petitioned the Federal Trade
Commission to request that the agency addresses the issue of social network privacy policies [60]. Moreover, some national data
protection commissioners have publicly warned of using certain
social network sites [62, 61]. Just recently, the vulnerability of
these services has been demonstrated by someone, who downloaded 100 million user profiles and made them publicly available
for download [63].

Joining groups within social networks can offer another exploit for potentially malicious de-anonymization attacks. A recent
paper [64] proved that 42 % of users that use groups can be
uniquely identified. These results are noteworthy, because traditional privacy attacks were based on aggregating information
from multiple datasets. Such methods were based on collaborative
filtering [65] and enabled an efficient and highly reliable character-
ization of a person from a few data. The underlying technology is
quickly advancing [66], and it may give service providers, such as
mobile phone, Internet television, or social gaming centers an un-
precedented amount of personal information. Research on related
privacy issues and their potential explicit or implicit consequences
is still in its infancy [67]. Moreover, efficient legal protection is
urgently needed. A simple-to-establish solution to some of the
above problems would be accountable pseudonyms [68].

Additional risks for the privacy of users emerge, when companies are forced to reveal private data to governments or legal
institutions. Google provides an example of the quantity of data
which is handed out to governments [69]. There are also joints
startup companies with the CIA [70]. Finally, when data are not
subpoenaed or stolen from cyber-criminals outside of the company, they can be leaked in the most fanciful ways, which go from

displacing a physical device containing sensitive information, to the dishonest action of a single employee from inside the company [71, 72].

11.1.5 Additional Ethical Concerns

Ethical problems are intrinsically "ambiguous, uncertain and prone to inevitable disagreement" [46], i.e. the correct answer cannot be deduced algorithmically from general rules to particular claims. They are related to cultural values and social norms. In the following, we raise a number of open ethical questions connected with large-scale data-mining activities, to which, of course, we cannot provide definite answers here. Related research programs are thus urgently needed. For the time being, governments and companies engaged in large-scale data-mining are advised to follow the procedural ethics approach presented in Sect. 11.1.6.

- As large-scale data-mining activities are increasingly successful in predicting (aspects of) individual behavior [73], they will constitute an extremely powerful tool. This raises issues of the possibility of misusing it. More importantly, it raises the question of who gets to use these tools on what grounds. Will it be national governments and international corporations? Would there be a moral imperative to make the systems available to developing countries, NGOs etc.?
- What about competing claims of systems? If an early warning system recommends certain activities, how should societies respond to such recommendations? For example, how to handle situations, in which a scarcity of resources occurs?
- Who will own the algorithms and the outcomes of the data-mining activities? Intellectual property is often discussed in terms of ownership of data used for input, but the more interesting question would seem to be: Who owns the predictions? As they could potentially be subject to patent protection for

computer programs and business methods, a rigorous analysis of the implications of intellectual property protection for data-mining activities is needed.

- If policy is based on predictions, how open is the system to critical review? Who will know and understand the algorithms? How can mistakes in algorithms be identified and rectified?

11.1.6   How to Address Ethical Issues in Large-Scale Social Data Mining

Large-scale data-mining raises both procedural and substantive ethical issues. Some of the latter are predictable and solvable by implementing legislative and technical solutions. In the case of privacy, for example, this would include

- the use of scanners for viruses, trojan horses, etc.,
- encryption,
- fragmentation of data [74],
- restriction of access/read/write/execution rights (depending on the type of data and purpose),
- selecting higher security standards in the browser (for example, turning off cookies or deleting the browser history),
- anonymous surfing [75, 76],
- use of pseudonyms.

Nonetheless, one needs to underline that a full understanding of substantive ethical issues would require a full knowledge of uses and applications of the system, which is impossible to acquire a priori. In order to ensure a future-oriented approach to ethics, every project performing large-scale data mining should therefore incorporate procedures that will allow the identification of substantive ethics as well as ways of addressing them. Such procedures should include the governance of the project from inception to delivery and cover governance recommendations for the individual components (early warning systems). It should incorporate reflexivity in the project team, continuously discussing the following questions (and regularly seeking independent feedback from outside):

- What are the substantive ethical issues that can be foreseen at any given point in time?
- What are the assumptions underlying the project itself as well as those underlying the ethical analysis (what is perceived to be an ethical issue, and why?)
- How can appropriate processes be established to address known ethical problems (e.g. informed consent procedures)?
- How can factual knowledge about the product and its likely consequences be gained?
- Who are the stakeholders affected by the system and how can their local knowledge be fed into the reflective process?

11.2 Towards Privacy-Preserving Data Analyses

Privacy concerns, although often justified, can cause serious obstacles to socio-economic data mining, while in many cases such data-mining would be in the public interest, when done in a privacy-respecting way. For example, socio-economic data mining would be needed to gain a better understanding of socio-economic problems, how they arise and how they can be addressed. Therefore, the following sections elaborate concepts that address how data mining could be done in a privacy-respecting way.

11.2.1   Deliberate Participation

The simplest possibility to do social data mining is to do it with data that individuals share deliberately. For example, some Web sites, such as Blippy.com, Skimble.com or Swipely.com, collect everything from consumer data over the last movie you have seen up how many push-ups you have done in your last training session. Participants of these Web services intentionally make their data available to everybody, and they can be analyzed in any possible way. The only concern from a statistical point of view is that the set of people participating in these Web2.0 activities is not representative for the entire population, i.e. one would need to make complementary analyses in order to learn, how it is possible to correct for biases in these data. Typically, participants are younger than average and are not concerned to share their data because they lead pretty much "average lives".

Further data can, in principle, be analyzed by crawling the Web. The data out there are usually traces of, for example, shopping activities at eBusiness platforms or social networking activities. They are accessible to everybody in small numbers, and it is not clear whether and how much people would care about a company or person analyzing these data in large amounts, as they can be gathered by automated programs such as "spiders". There are certainly problematic applications of this kind, in particular when the resulting datasets are used for business purposes, although the data were not intended for this, or if they are sold to third parties with unknown intentions.

As the recent discussions about the activities of large data-mining companies shows, legal regulations against unauthorized processing of individual data are urgently required. Scientific analyses, which lead to discoveries of public interest, may have a better justification, but it must nevertheless be decided in each single case, whether individual rights are touched and what is the public benefit of such analyses. Shear curiosity and the publication of a scientific paper may not be a sufficient justification, and therefore, the consultation of an ethical committee seems appropriate.

As a consequence, it would be much better to work with data that people provide intentionally for a given purpose. Statistical samples can already be quite useful. Special "on-demand-data-gathering" tools could allow people to easily opt-in and opt-out of data-collection programs in a situation-specific way. For example, while people may usually object to provide their data, it is likely that the participation rate increases in special situations such as crises, where people tend to change their priorities and make a contribution. However, it is fundamental that the gathered data will be used only for the purpose people have explicitly given consent to. With the project "Gaydar"[77], the MIT demonstrated how easy it is to filter out sensitive personal information, which may be misused, from publicly available data. This study predicted the sexual orientation of Facebook users by analyzing the publicly accessible pictures of their friends. Such studies suggest that the processing of data should be allowed only for a certain time period and for the purpose they have been provided, requiring that users have adhered to an explicit, fair, and informed opt-in procedure. For sensitive data-mining activities it would be appropriate to apply the standards followed in clinical studies today. In order to support on-demand participation, particular trust-worthy Internet platforms should enable the case-wise sharing of personal data according to the specified purposes. This could be a special function of future eGovernance platforms.

11.2.2   Anonymization and Randomization

To satisfy the data protection directive 95/46/EC, any data containing personal information needs to be anonymized before it is evaluated. While this may be sufficient for many simple analyses, it may not guarantee that the identity of individuals cannot be revealed from anonymized datasets. Substantial research has been and is currently being performed in the database community on privacy preserving data mining, reflecting the importance of this subject [78, 79, 80, 81, 82] (for a comprehensive state-of-the-art summary see the "Privacy-Preserving Data Publishing" survey [82]). Nevertheless, there are still a number of open problems, and many approaches have been made that all lack user-friendliness, integration, and a consequent systemic approach. Problems occur in particular when datasets contain a list of many different features, and some combinations of features are rare. As a consequence, such data must be sufficiently coarse-grained and/or randomized to make sure that combinations of features occur in sufficiently large numbers and cannot be individually resolved. Furthermore, one must avoid to save lists with many features in one single dataset. It is safer to store them separately on different computers and to access the separate datasets only with programs which are guaranteed to determine coarse-grained properties only such as (sufficiently rough) statistical distributions. The resulting derivative datasets should be comparatively small and unspecific, or they should be surrogate datasets, in which the relevant *statistical* properties are the same, but the underlying individuals (persons, companies, etc.) are randomly reshuffled and not identifiable.

The generation of the anonymized, derived and surrogate datasets for the original data should be done by particularly qualified and trustable institutions, while a larger number of people can work with the resulting, less critical datasets. In the last decade, research in privacy-preserving data analyses has produced methods and tools aimed at publishing data under a privacy-preserving shield. For example, data are made anonymous with respect to a certified trustable anonymity notion, which essentially guarantees that the probability of tracing back any data to the identity of the person to whom the data originally belongs is so low that it can be considered null in practice. Another active research line concerns the privacy issues in case of mobility data such as those produced by location-aware devices [83, 84].

To protect the original datasets from theft and unauthorized access, the specially secured and authorized data centers should store them in an encrypted format, and decryption should be done only piecewise and for the milliseconds, when the derivative data are generated. All commands and source codes of computer programs involved in sensitive operations should be automatically logged on a separate server, which is unaccessible to persons who are authorized to deal with original datasets.

11.2.3   Coarse-Graining, Hierarchical Sampling, and Recommender Systems

As indicated before, in case of sensitive data (such as pregnancy, religious confession, diseases or the sexual orientation), it must be ensured that individuals and group memberships cannot be identified from socio-economic datasets. For this reason, datasets for statistical analyses must be coarse-grained in a suitable way. This may also be done by real-time data-mining ("reality mining") approaches, if they are suitably designed. For example, to determine congestion on a freeway, it is possible to analyze mobile phone usage data, but it is not at all necessary to know who is calling whom and what is the content. The same applies to GPS localization information of mobile phones, if the distribution of people is determined for the sake of an efficient evacuation. It is just necessary to make sure that any potentially sensitive data (such as the underlying phone number) is deleted before the statistical evaluation is performed. However, as the recent case of WiFi recordings by Street View cars has shown, transparency is needed for such applications, as one needs to make sure that no sensitive data are stored. In principle, it could be legally required that the underlying algorithms are published, and no algorithms may be used which are not open source.

One particular approach in reality-mining could be a hierarchical sampling via ad-hoc networks of, for example, sensors or mobile phones, where detailed information is only processed locally, and any transmitted information undergoes a certain level of aggregation. That is, as data are distributed over larger distances, they undergo several aggregation steps, which may be imagined like a hierarchical sampling method. Whoever wants to process a large dataset, would only get a coarse-grained view of the data, since they would be accessible only via a high level in the data-processing hierarchy. Whoever managed to see data on a lower and, therefore, detailed level, would only have a very short-sighted and limited view, i.e. see very little. It appears, however, that the technical details of such systems matter in order to be sufficiently privacy-protecting and acceptable to users of the resulting services (e.g. location-based ones). A transparency of the data-processing algorithms and related legal regulations appear to be needed. It should be explicitly forbidden and prevented to collect and store low-aggregation-level data. It must be ensured that they are deleted directly after they have been processed and before they are transmitted. To be uncritical and widely acceptable, the processing should happen in the technical devices used by the individuals and not on company-owned infrastructures (as is the case nowadays).

A possibility to make low-level data robust to interception would be as follows: Given that the data of interest can be represented as points in a (quasi-)continuous space, one could add random numbers according to a certain statistical distribution. Rather than transmitting the correct value (such as the exact location of the individual), a random number ("noise") would be added, before the value is transmitted to the ad-hoc network performing the reality mining. Such random falsification would make low-level-aggregated data useless and create a "foggy" situation that protects the individual from being revealed [85].

However, if done in a suitable way, the aggregation of the individual data could still lead to reasonably accurate results due to the law of large numbers, according to which errors average out in a statistical sense.

Services of *recommender systems,* of course, need to target an individual specifically, which seems incompatible with overlaying noise. However, recommender systems could still be realized by applying a two-component strategy: The first component would be a rough search, which does not consider individual information or preferences (or only, when sufficient noise is overlayed). Among the search results, the personal computer or smart-phone of the user would then select the individually fitting search hits, products, or advertisements, based on personal information and preferences that are exclusively stored on the individual computer rather than on a system of servers. Putting it differently, recommender systems should be changed from an approach, where individually customized recommendations are pushed to the user, to a pull approach, where the user selects in confidence one option out of a larger spectrum of downloaded recommendations in a way that does not reveal his or her preferences. Individuals who are even concerned about storing personal information and preference data on their own computational device should have the possibility to turn off the second component, which would then result in untargeted research results and in recommendations, which would not be individually customized.

The same approach can be used in connection with location-based services, the great comfort of which many people do not want to miss anymore. Let us assume somebody wants to be guided to an erotic shop, but does not want the guiding company to know this. The person would go to the center of town, and based on his or her falsified, approximate position, the GPS location service would forward to the mobile phone information about shops in the area. The user could then select among these according to categories, but the selection would only be known to his or her phone. It would not be forwarded to the content provider, nor would the exact location of the user be known to the provider.

11.2.4 Multiplayer Online Games, Pseudonyms, and Virtual Identities

Another possibility to study social interactions are offered by multiplayer online games such as Second Life. The advantage of these games is that players can participate under pseudonyms, without revealing their real identity. From an experimental point of view, this has some side effects, as people may behave differently under anonymous conditions as compared to conditions with face-to-face interactions. Still, these effects may be compensated for, and there are a number of behaviors which occur quite realistically. For such reasons, studying interactions in multiplayer online games is becoming a research technique, which is used complementary to lab and Web experiments [86, 87].

Some of the artifacts of studying multiplayer online games result from the following facts (here we assume that the system would not allow the registration of several *identical* pseudonyms):

1. People may change identities, i.e. register as a new user if their previous behavior is sanctioned by other players or by the system ("whitewashing").
2. People may use multiple identities, potentially also in parallel.

To overcome these problems, the following measures can be taken:

- Everybody could get a unique virtual identity, which would be needed to create unique pseudonyms.
- Registering a new identity could be made very time-consuming or costly.

- People may be allowed to join a multiplayer online community by invitation only (and there would be separate lists of members and pseudonyms, which would be secret and encrypted).

An additional problem is that people may buy an identity (pseudonym) with high reputation or scores from somebody else. This problem may be addressed by performing behavioral consistency checks to reveal the use of the same identity by different people. Alternatively or complementary, the matching of pseudonyms with the unique virtual identity could be sporadically checked (by requiring to enter it).

A unique virtual identity can be generated by a trustable public institution such as the registration office. It is practically an electronic signature that can be used to submit documents such as tax declarations or payments. Note that there are already private companies offering trusted virtual identities/electronic signatures, among which Verisign, GeoTrust and Thawte.

The unique virtual identity would have a finite validity (i.e. it would have to be regularly renewed), and plausibility checks for identity thefts would be made, to invalidate stolen identities (such as for credit or debit cards). The identities could, for example, be generated as follows (where the system would log which administrative person handed out what card): When asking for the virtual identity, a box would be ticked in the files of the respective person, indicating that a unique identity card has been handed out. The identification number of this card would be randomly generated, and the receipt of this card would be confirmed with a signature, showing a valid photo document (ID or passport). The identification number, however, would *not* be known to the office handing out the card.

To reveal the real identity behind a virtual identity in case of a severe crime, this should require the simultaneous agreement of several independent authorities (e.g. judges [who could also be from trusted non-profit organizations]). Only by combining the keys of two thirds of the respective responsible authorities, it would be possible to reveal the real identity. This can be ensured by saving bits of the identity code in different databases, all in an encrypted way. The access keys and responsible persons controlling these keys would be regularly replaced by new ones in order to avoid corruption. It would also be good to let computers randomly decide, which ones of a number of authorized persons would have to decide whether to reveal the identity or not. This would minimize external influence on the decisions of the respective responsible authorities.

11.2.5   Anonymous Lab Experiments

Social behavior can also be studied in lab experiments [31]. In these experiments, one may want to ensure anonymity of the participants, as they may otherwise not reveal their true opinions or their normal behaviors. Such experiments may require that the experimental subjects do not meet the experimenter, and maybe not even meet other experimental subjects.

There are different ways of implementing such a design. For example, individuals randomly passing by an information stand could be invited to participate in the experiment. If they were willing to participate, they would draw a lot with a unique number, and they would enter the number of the lot into a time table, which is hidden from the experimenter. At the time of the experiment, the experimental subjects would show up in separate rooms, where they take their decisions in isolation, based only on information coming from a computer screen. Their decisions would then be transferred via Internet to the other experimental subjects. After the experiment, the experimental subjects would receive an envelope with their compensation, which would be pushed into their rooms through small slits under their locked room doors. The subjects could leave their rooms 2 min later.

The experimental setup would ensure that nobody would know, who participated in the experiments, and it would be unlikely that the same person would participate several times. Nevertheless, participants may be suspicious whether this setup will really be performed in an anonymous way or whether there is a chance of hidden observation, and this may still affect their behavior.

A similar and even more privacy-protecting setup can be realized via a Web experiment. A large number of people would be informed that the experiment takes place at a certain time and could log on with pseudonyms. Among the people who have logged on the experimental webpage, the computer would randomly match individuals to form experimental groups. At the end of the experiment, each individual would get a voucher with a unique code, which can be exchanged for the compensation for participating in the experiment. One of the following ways of payment may be chosen (listed in increasing order of anonymity):

1. The experimental subject gets the compensation from an independent cashier (e.g. the university cashier) by showing the voucher, without the need to sign a receipt.
2. The person gets the money from an independent, i.e. trusted third-party payment service (e.g. bank or post), when presenting the voucher (i.e. the voucher would basically be a cheque). For example, it would be possible to use the mechanical turk [88] for a third-party payment.
3. The experimental subject gets the compensation by entering the unique code of the voucher into a special cash machine.

Experimental subjects could be recruited in different ways: The simplest would be to display posters in public areas, calling for participation at a specified time via a certain Web page (and people could actually participate from a computer in a computer pool or Internet cafe, if this gives them a better feeling of anonymity). Similarly, the announcement could be made via an advertisement on a heavily frequented Web portal. At the specified time, an algorithm would match visitors of the website in a random way and try to make sure that groups of friends could not play with each other.

Avoiding that certain subjects participate in the same experiment multiple times is more difficult (at least as long as only a few people have unique virtual identities). One possibility would be to send out invitations to a large number of e-mails, making sure that there is only one e-mail address per person. People willing to participate would enter into a webpage their e-mail address or a unique code sent with the invitation e-mail. This is required for authorization, to prevent multiple access. After this, they would be redirected to a webpage, which shows a large list of unique codes, one of which can be randomly chosen by clicking on it. This will cancel it from the list and inform a web service hosted by an independent, third party (e.g. a computer center) that this code has been authorized. When the participant enters the code into the Web page of the independent Web service, another code is returned, which is randomly selected from a long list of unique codes. That code will be needed to get access to the experimental platform at a later point.

The above procedure makes sure that the first step prevents multiple access. Afterwards, the selection of an individual code makes sure that the third party cannot have any clue of the relation between this chosen code and the e-mail address of the experimental subject. While it knows the list of acceptable codes, it does not know the identity of the person, just the fact that it is authorized to get a randomly chosen code from a list of unique codes, which are accepted by the experimenter. However, which code is randomly selected by the computer of the third party cannot be known by the experimenter. Finally, any temporal correlation among individual registrations is lost by implementing a sufficient time delay, after which the actual Web experiment takes place.

11.3 Concept of a Future, Self-organizing and Trusted Web

In the following, we will describe technologies, which give people back control over the data available about them. Some of the following runs under the label of privacy enhancing technologies (PET). For example, most Web browsers today allow one to turn off cookies (which, however, makes certain Web services disfunctional). Furthermore, there are tools such as Tor [75] and Freenet [76], which support anonymous Web browsing and anonymized content sharing by obfuscating the IP address of a computer. However, this is still not sufficient to guarantee anonymous web browsing [57, 89]. Furthermore, one serious problem of today's Internet still is the fact that it does not forget and that it does not provide control over copies of data, which somebody has uploaded in the past (e.g. party photos). First solutions for data with finite lifetimes have become available only very recently [90].

11.3.1 Data Format

The following concept of a future, self-organizing and trusted Web is aimed at overcoming the above-mentioned and other problems. The basic feature of the concept is a new "Helbietti" file format, which electronically signs and encrypts contents, but has a number of unencrypted specifiers such as

1. a unique file identifier (which is different for copies),
2. the kind of content (factual information, advertisement, opinion, unspecified),
3. the lifetime (from ... to ...),
4. a public annotation field allowing to tag the file and link it to others or to link it to comments or ratings, and

5. information regarding the price of producing and receiving a copy of the file.

There would also be encrypted specifiers (readable only to authorized users), such as

- the originator of the data (anonymous, pseudonym, real name, or company name),
- the owner (anonymous, pseudonym, real name, or company name),
- the date and time of generation,
- a unique content identifier (e.g. check sum),
- locations of authorized copies, and
- the persons or groups authorized to read, modify, or execute the file (which would again be based on real names or pseudonyms etc., but one possible specification would be "everybody").
- Annotations, which could be read only by the authorized persons or groups.

The following data would be double encrypted and accessible only to the owner of the file (and jointly to a specially authorized group of inspectors, see below):

- the file identifyer(s) of the file(s) it has been derived from (i.e. the previous version(s), if one existed, otherwise null) and the files that have been derived from it (e.g. any identical or modified copies),
- all information regarding money transfers between customers or users of a file and the owner of its content as well as the respective tax authority, and
- the digital rights management settings (e.g. maximum number of copies that can be made from the original file).

To ensure that privacy and intellectual property rights are not undermined, checksum error-detection techniques would immediately reveal unauthorized manipulations to the original copy.

Semi-automatic filtering measures could be implemented on servers which would refuse storage and forwarding of tampered copies. This kind of filtering may be compared to the immune defense system of the body against harmful viruses etc. For issues of copyright protection see Sects. 11.3.2 and 11.3. Moreover, depending on the sensitivity of the data (public, restricted, confidential, secret, etc.), they would be fragmented and distributed over several files stored in different locations [74]) and additionally password-protected, potentially requiring several passwords from independent authorized persons to access them.

11.3.1.1 Finite Lifetime Data that can be Controlled

This concept immediately allows one to limit the lifetime of data, as they could only be decrypted within the specified time period. (Although there are first software solutions in this direction [91, 90], they seem to require further enhancements.) In order to avoid tricking the file by modifying the time on a particular computer, the file would automatically have to verify the time with one or several randomly chosen, trusted servers (depending on the level of confidentiality; of course, there would be a long list of such servers). Additionally, the file could be opened in this time window only by individuals or groups that are listed as authorized.

Besides, one could foresee a further restriction to the access of a file by requiring that either the original file or one of the authorized copies are still accessible somewhere in the Internet. That is, if the owner of the file would delete the original file and the authorized copies he or she may have created as backups, no copies of the file may be opened any longer. This would give the owner of the file perfect control over its distribution—a fact which is also important for copyright protection (see Sect. 11.3.2).

11.3.2   Intellectual Property Rights

The new data format also provides new possibilities to protect copyrights better. As music or video files would be encoded and require a certain password to be opened, access to the file could be restricted to a single user or group of users. Moreover, Helbietti-formatted files could be set up in a way that a certain prize is charged (e.g. to a prepaid account) whenever a copy is produced. During the copy process, this amount of money would automatically be transferred to the owner of the intellectual property rights, the intermediate seller (e.g. a shop or the person whose file is copied) and the respective tax authorities. This would facilitate a "viral marketing", where users are distributors, who can earn money by disseminating file contents, while benefiting the owner of the intellectual property rights. This would, of course, not prevent the recording of videos and the illegal distribution of related copies. However, this problem could be minimized by a combination of the following measures:

- using pricing schemes that people consider fair,
- selling copies of different quality at different prices,
- allowing users to download contents with pseudonyms and anonymous payment services (e.g. [23]), such that providers cannot track which contents are bought by what customers.

Massive copyright violations could be reduced by using the labeling, reputation and sanctioning mechanisms described in Sect. 11.3. Also note that the proposed file format allows one to make all copies unaccessible by deleting the single file that the copies were derived from (see Sect. 11.3.1.1). Finally, for serious cases of piracy the new file format provides a possibility to track from what file a copy was derived, if decryption has been decided by a number of specially authorized people (see Sect. 11.3.3).

11.3.3   Trust Management

11.3.3.1   Rating and Reputation

Many public goods such as reliable information systems are very difficult to create and easy to exploit and/or destroy. This creates dangers for the quality of "the commons" (public goods). In absence of clear responsibilities, such as it is often the case in the Internet, large collaborative efforts are not encouraged. In fact, contributors are more difficult to identify and to reward, while vandals and other detractors can easily thrive.

Therefore, the self-control of the Web, based on suitable reputation concepts, would be a desirable feature. In principle, people should be able to rate, tag and comment on any data they have accessed. Also ratings and comments could be rated, which would earn the rater a certain reputation. Ratings would not be one-dimensional, but done on a multi-dimensional scale (which could be customized in a user-specific way). The multi-dimensionality is important to support pluralistic, community-specific views.

Note that details of the design of the rating mechanism are crucial. Manipulations of ratings must be prevented. The rating of the raters can serve this purpose, if well constructed. It determines their weight in the calculation of an aggregate rating. The design should be able to distinguish votes coming from robots and from humans. Furthermore, whitewashing (a new pseudonym) and sybil attacks (the creation of many pseudonyms) should be prevented (see the previous section regarding possible ways to do this). Furthermore, to disclose a manipulation of the own reputation via pseudonyms one is controlling (or a mutual manipulation through a friendship network), consistency checks will be needed. That is, at random times, it will be necessary to compare the reputation values that a pseudonym has from the point of view of several others (randomly chosen interaction partners, also independent outsiders). This comparison of reputation values is something like a gossip strategy. If the values are sufficiently consistent, everything is fine and the reputation seems reliable. Otherwise, there are reasons to be suspicious. In such a case, the pseudonym would be labeled for the purpose of intensified observation in order to reveal the manipulation. Such a differentiated inspection strategy, which focuses on indviduals with a suspicious reputation (and newcomers), but which otherwise restricts to random inspections, saves computational resources but can reduce the level of fraud.

Furthermore, the contents that users upload in the Internet would be rated by other users who have access to them, earning the provider of the content a certain reputation. This offers a tool to separate high-quality from low-quality contents. In order to avoid opinion dictatorship by the majority and ensure socio-diversity (pluralism), it will be necessary to allow for community-specific and multi-criterial ratings. Communities would either result from social networks, or they could be determined via community detection algorithms, identifying groups of people with similar rating, tagging, and commenting habits, i.e. with similar preferences and tastes (so-called "quality collectives" [92]). It should be remembered here that the identities of the people belonging to a community will usually not be known, but rather be composed of virtual identities, namely when pseudonyms are used.

The community-specific ratings, tagging and comments can serve to create filters for certain contents. Therefore, it is possible to design community-specific recommender systems which prioritize contents fitting a community's or an individual's taste. Similarly, undesired contents can be excluded so that it becomes possible, for example, to protect children from sexually explicit or violent contents. In other words, users could tag illegal or inappropriate contents. In serious cases, this could trigger sanctions (see below) or even legal action. For instance, the access to the file could be restricted (e.g. to people above a certain age), or the decryption could be disabled by a certain code foreseen by the cryptographic algorithm. Also, in case there is evidence that access to certain encrypted contents is in the public interest (e.g. relevant for public security), the encryption method could foresee a decryption. In order to avoid misuse such as censorship or violation of privacy, both, the decision to restrict access and to enforce decryption of a file or list of files, would need a certain number of randomly selected, generally trusted and authorized people to agree on the action that needs to be taken. Consequently, such actions would require the application of several keys at the same time. To avoid unjustified decryption by bribing authorized people, these should be replaced after a certain time period, which means that the keys unlocking a file need to change or be changed over time.

11.3.3.2 Sanctioning Mechanisms

In reality, a reputation is hard to earn, but easy to lose. This suggests that, besides a reputation mechanism, the self-organizing Internet could foresee certain sanctioning mechanisms to facilitate a high level of quality. Sanctions may include everything from low ratings, over certain kinds of tags and critical comments, up to banning specific contents within a certain user community. Particularly destructive behavior may be sanctioned by temporary bandwidth reduction. For instance, manipulating ratings or reputation values by sybil attacks (self-ratings via multiple pseudonyms) should be sanctioned in one way or another. The same applies to wrong declarations (e.g. labeling advertisements or opinions as information). People should be free to express their opinions, but they also need to have a chance to distinguish opinions from facts. Furthermore, spamming the Internet with low-quality contents should be sanctioned. Note, however, that what constitutes low-quality content for one community could constitute high-quality content for another community. That is, the sanctioning would usually be community-specific. Only in exceptional cases would it be generally applied.

11.3.4   Microcredits and Micropayments

For the future of the Internet it also seems wise to consider the possibility of collecting microcredits for small contributions to the public good "Internet". Such microcredits would allow one to reward people, for example, for contributions to public encyclopedias or also for rating contributions or reviewing (commenting on) them.

The data format of microcredits would, therefore, not only contain a certain value ("number of points"). It would potentially also contain (usually in a sufficiently anonymized or encrypted way) information about who owns it and what it was earned for or paid for. Moreover, it would be a tradeable unit, which could contain pointers to who owned it last and whom it is being paid to (again in an encrypted way). Having both backward and forward pointers supports double book-keeping when needed. In mathematical terms, rather than a being scalar (which implies a number of fundamental problems), a microcredit would be an element of a microcredit network connecting values with pseudonyms and merits or items bought. These elements would have a certain number of weighted links (in-degrees and out-degrees) reflecting cash flows. Therefore, it would be possible, in principle, to distinguish different kinds of currencies for different kinds of contributions, and it would also be feasible to a certain extent to analyze flows of microcredits between pseudonyms over time in a privacy-respecting and confidentiality-protecting way (see the section on reality mining regarding how to do this; note that companies could use different pseudonyms for different organizational units, and that they may change them over time). Such kinds of analyses would be enormously useful to determine instabilities in the microcredit market. It would also be possible to give money a history and, therefore, distinguish "dirty money" (such as "blood diamonds") from ethical investments, as certain customers demand them today.

11.3.5 Transparent Terms of Service

In order to sign-up to a service in the Internet, one is more and more often placed in front of a long list of obligations and contractual clauses applicable for any sort of special case, for which an ordinary user does not have the adequate competency nor the necessary time to understand. The result is that they are skipped and blindly accepted. Based on such "acceptance", companies can grant themselves a great freedom of action in handling the personal data of their users. This should not be allowed, and large-scale data-mining activities should be protocoled and publicly controlled.

Anybody willing to start collecting data from the Internet, or other private and public nets, should first publicly provide a legally binding declaration about what is done with the data and why. In particular, it should contain whom (what companies, institutions, etc.) the data will be shared with, and what is done with them exactly.

This declaration should also comply with an international *"data-collecting protocol"*, which needs to be established to set legal and ethical constraints to the action of *data harvesters*. The protocol should define minimum quality of service standards, e.g. regarding waiting times of customer services, times to delete private data, fees, how to contact the data management center's service, whom to contact in case of complaints.

Compliance to the protocol would allow companies to show a "Privacy-Safe Badge" on their website, which would immediately be recognized by surfers (see [93, 94]). Showing the privacy badge would probably become fundamental for certain categories of companies operating the Net (e.g. search engines, social networks, banks, etc.). Not having the badge, could make a relevant difference in the trust level of customers. Moreover, it is easily foreseeable, as data-mining activities become more pervasive in the future, that the importance of the badge would eventually extend to other general purpose websites.

The badge would be granted by newly created (ideally supernational) rating institutions, which should also have the authority to enforce the standards related to the respective security badge by inspection. Such an institution will be the only legal parties empowered to issue privacy badges and revoke them in case of misconduct. In future, the collection of personal data on the Internet without a proper badge could be considered an illicit activity and insofar be sanctioned by users accordingly.

To obtain the badge, interested parties would have to demonstrate that they possess both, the *ethical* and *technical* standards necessary to accomplish such a delicate data mining task. After proper checking, and depending on the purpose of the data collection specified in the harvesting declaration, different types of badges could be issued. Each badge would also be linked to a standardized user licence.

In order to add dynamism and a more democratic taste, the badge could foresee user ratings and comments. These opinion feedbacks per se, would not generate legal consequences for the owner of the badge, but would help to detect misconducts earlier and to alarm the community, and it would trigger inspection procedures by the issuer of the badge.

Finally, users should be able to a-priori set their preferences and conditions on their browsing devices, under which they are ready to participate in data-collection campaigns or not[2]. Browsers

---

[2] For example, willingness to allow collection of personal data only for scientific purposes.

would immediately examine the badge of any visited website, comparing it with its stored preferences and automatically notify any threat to the privacy of the users. Besides, this would solve the notorious issues of unreadable or over-technical "Terms of Services" conditions, which should not any longer be read directly by users themselves.

11.3.6 Privacy-Respecting Social Networks

Social networking has been rapidly spreading in the past years despite frequent warnings regarding a lack in privacy protection. Recently, for example, somebody succeeded with downloading 100 million user profiles and uploading the dataset for free use by everybody [63]. It is often claimed that users simply do not seem to care about uploading private information to the Internet. However,

- this does by far not apply to everybody (in fact, most computer users still do not have social networking profiles),
- the terms of use have changed since most of them have uploaded their private data (e.g. photographs),
- some users may assess the comfort of the service provided by social networking sites higher than the current side effects, but this may change over time.

Besides, a recent empirical study has impressively demonstrated that people *do* care about the use of their activity data [95].

It certainly appears necessary to have alternative technical solutions for social networking, which protect privacy better. A first project of this kind is DIASPORA [96], which intends to decentralize the storage of sensitive information.

Privacy-protecting social networks could be imagined as follows: Individuals would only see part of the network. Individuals and communities could determine what can be seen to outsiders of the community and to whom (friends, friends of friends, second-next-nearest neighbors, or everyone; same with business partners). Depending on this, certain kinds of information would not be visible to outsiders, others would be (as communities may want to gain new members). In essence, surfing in social networks would be like travelling between communities, and this would feel like visiting other countries. While certain things would be visible, others (the private part of the information) would remain hidden to strangers (as private houses are).

11.3.7 Summary

In essence, many of the problems of the Internet today result from Web2.0 and other applications, which the Internet was originally not designed for. Consequently, current technical solutions are insufficient. A new way of organizing the Internet appears to be needed and possible. Suitable solutions can be developed by transferring concepts of social self-organization to the design of the future Internet. This constitutes an interesting challenge within the research field of techno-social systems.

11.4   Recommended Legal Regulations

Currently, data about people are probably processed, used and misused in any conceivable way. Since regulations are insufficient and heterogeneous, the situation has sometimes been paraphrased as Wild Wild Web. It is therefore not surprising that the EU Fundamental Rights and Citizenship Commissioner Viviane Reding has recently pointed out [97] that Europe needs more harmonization regarding a data protection law. Determining the best routes towards this goal deserves targeted research. However, as the problems are acute, action needs to be taken soon. Therefore, the following sections make a number of suggestions.

Given the problems of the current Internet and the foreseeable future developments, data collection for research or for business should be regulated taking into account privacy, legal requirements, science's and business' interests. We foresee that methods of data collection should be open, controllable and verifiable by legal authorities and the public. Legal procedures and the law should establish what type of data can be collected, what type of data may not be collected, and how the sensitive part of the collected data must be hidden from people or organizations collecting them at each point of the data collection procedure. Methods of warranting the safety of sensitive data should be public and should be verifiable at all times before, during, and after collection. For example, we recommend to work out legal regulations for the following:

- Data storage, access, processing and usage standards should be fixed for public, commercial and private entities operating in a certain country. Transparency regarding the storage, access, processing and use of data should be enforced. In particular, there should be a binding public declaration of what kind of private data are being stored, processed and used, and how this is done.

- Personal data should always be stored in an encrypted way. However, it should be made easy to inform oneself free of charge about the data determined and stored by other individuals, companies or institutions, and how these data are accessed, protected and used. Therefore, technical solutions should be required, which allow individuals to access (and decrypt) their personal data online and to delete data one does not want to be stored (if there is not a law that requires such storage). Furthermore, it should be easy to opt out from the determination, storage and/or processing of certain kinds of data. It should be possible to sanction violations of this right efficiently, and affected individuals should be properly compensated.

- One should establish standards ensuring informed consent of users with the data an information system is determining, storing, or processing. Users should not be forced to agree with a storage, processing and use of information that is not technically required for the services a user wants to get. For example, providers of media contents should not force customers to reveal their identities (and effectively their preferences via the contents they buy), as the contents or services can also be paid for anonymously. Putting it differently, companies should be required to offer, in a clearly marked way, options to customers that allow them to choose at any time between a data-rich variant (providing the service provider with many detailed individual data) and a data-poor (privacy-protecting) variant without artificially created disadvantages (which would effectively force customers to reveal their data). Within fair limits, it would be acceptable though to charge a higher price for data-rich services to users, who have chosen the data-poor variant themselves.

- Licence and usage agreements of software products and information services should be regulated and standardized. As most users do not read or understand the terms of use, and as they do not have any chance to negotiate these conditions, there should be a few (certainly less than ten) standard licences, which should be indicated by a color or other codes, whenever a certain software or information-based service is used. Alternatively, softwares and services should be rated by independent agencies based on the benefits users can expect from them and the degree to which privacy and confidentiality are potentially affected.

- It would be useful to define the individual and corporate responsibilities for damages created in the virtual or real world by activities in the Internet.

- However, considering the fact that the content of a file is revealed only when it is accessed, it should not be possible to punish people for the access of contents, if the contents are not warned of in advance in a sufficient and qualified way (e.g. based on the rating and reputation system suggested above). In other words, users should be protected from legal traps in the Internet.

- Considering the abundance of free contents in the Internet, it is advised to implement a copyright, which considers the facts of modern information systems and requires copyright holders to make proper attempts to protect their products from unauthorized access (e.g. to indicate their copyrights, encode electronic files, and offer simple, fair, and anonymous payment procedures).

- Compensations for privacy violations would have to be fixed, and legal procedures would need to be simple and effective to allow people to protect their rights. For example, fines to companies, which sell private data without authorization, should be significantly higher than the likely profit they can make on such business.

- The priorities in cases of conflicts of interest should be worked out clearly (protection of individual human rights comes before collective public interests, which comes before institutional interests of companies or political parties, which comes before individual interests).

- Legal regulations should protect individuals against discrimination based on private data and guarantee an efficient compensation in case of violations.

- The introduction of class action would allow users to better defend their rights at court against individuals, companies, or institutions violating them, but the implementation should consider that the way attorneys are compensated and the way discovery is organized in civil procedures largely determines how desirable and effective class action lawsuits are.

- Unique virtual identities/electronic signatures should be offered for everybody.
- It should be required to specially mark Web links that are redirected to contents with a different character, or Internet services that are changing their character (e.g. from non-commercial to commercial), or the use of pseudonyms that have been used before by others.
- It may be useful to fix a statute of limitations, i.e. a time period after which violations of Internet-related regulations can no longer be sued. These time periods should increase with the seriousness of the violation and its consequences. It should also depend on whether the effect of the violation was in the past or relevant for the presence and future as well.
- There should be legal procedures regarding the random and targeted control of the fulfilment of legal standards regarding the storage, access, processing and use of private data.
- Conditions should be worked out for imposing access restrictions or forced decryptions of suspicious Internet contents, in case there is evidence that they seriously threaten the public security (such as instructions how to build bombs). Such measures, their extent, and results would have to be reported to the public, and individuals would have to be compensated, if it turned out that they were unjustified.
- Companies receiving public money should be required to make data of public interest available for research, after they have been processed in a way that removes sensitive information (see the above sections on how this can be done).
- It would be good to have neutral, publicly controlled third-party infrastructures, which allow to perform anonymous Web experiments and data mining.
- Special procedures should be defined for cases, where access to original or sensitive anonymized data is justified and required (e.g. for certain kinds of research of public interest). A good example is the way in which Harvard University regulates the

access and processing of the data of the Framingham Heart Study, which allowed scientists to discover social processes promoting the spreading of obesity, smoking, depression, or happiness, to mention just a few relevant examples of gained insights that can be beneficial for individuals and the public [98].

- There should be a fair right of information and participation in social activities mediated by ICT systems. For this reason, information businesses directed at a mass audience and with a large market share should be required not to discriminate and exclude certain user groups through inappropriate pricing schemes or terms of use (e.g. the requirement to agree with the arbitrary use or transfer of personal data or the requirement to allow for cookies, where this is technically not needed to provide the requested service). Individuals should always have the possibility to opt out of data uses they do not agree with, without losing access to information services not requiring these data.

11.5 Recommended Infrastructures and Institutions

In order to have a powerful, largely self-regulating Internet, the following kinds of institutions would be useful to have:

1. Public data centers, which perform a neutral and independent data collection that is not driven by the need to make money, but serves the purpose to inform the public in the best possible way. Such a system could implement the reputation, community formation, sanctioning and privacy respecting mechanisms discussed before in connection with the concept of a self-organizing Internet.

2. Research centers, which study what can be done with publicly available data, to assess the potentials and risks. These centers should also develop the technology of the self-organizing Internet sketched above.

3. Publicly controlled, neutral institutions, which can serve as independent third parties in experimental designs that ensure anonymity (see Sect. 11.2.5).

4. Independent quality audit centers, which evaluate the level to which companies protect privacy and provide good services and fair terms of use.

5. One or several complaint center(s), which collects complaints of Internet users and can take action against illegal or unethical practices. These centers should be well connected with the public media.

6. An ethical committee, which assesses risks of information technologies and markets. It should set ethical standards regarding the storage and processing of data and support the preparation of required legal regulations.

7. A center working out contingency plans for the case of large-scale failures of information and communication infrastructures, e.g. due to denial of service attacks, spam, viruses, trojan horses, worm or phishing problems, or solar-storm-related failures of electronic systems.

8. A committee working out suggestions for legal settings, as the need for institutional regulations arises through new technological developments.

11.6 Summary

Socio-economic data mining has a great potential in terms of gaining a better understanding of problems that our economy and society are facing, such as financial instability, shortages of resources, or conflicts. Without large-scale data mining, progress in these areas seems hard or impossible. Therefore, a suitable, distributed data-mining infrastructure and research center should be built in Europe.

Reality mining provides the chance to adapt more quickly and more accurately to changing situations. For example, it will facilitate a real-time management of challenges like evacuation scenarios or economic stimulus programs. Further opportunities arise by individually customized services, which however should be provided in a privacy-respecting way. This requires the development of novel ICT (such as a self-organizing Internet), but most likely new legal regulations and suitable institutions as well.

As long as such regulations are lacking on a world-wide scale (and potentially even thereafter), it is in the public interest that scientists explore what can be done (in a positive and negative sense) with the huge data available about virtually everybody and everything. Big data do have the potential to change or even threaten democratic societies. The same applies to sudden and large-scale failures of ICT systems. Therefore, dealing with data must be done with a large degree of responsibility and care. Self-interests of individuals, companies or institutions have limits, where the public interest is affected, and public interest is not a sufficient justification to violate human rights of individuals. Privacy is a high good, as confidentiality is, and damaging it would have serious side effects for society.

Acknowledgements The authors of this White Paper are grateful to Karl Aberer, Andras Lörincz, Panos Argyrakis, Endre Bangerter, Andrea Bassi, Stefan Bechtold, Bernd Carsten Stahl, Rui Carvalho, Markus Christen, Mario J. Gaspar da Silva, Fosca Giannotti, Aki-Hiro Sato, David-Olivier Jaquet-Chiffelle, Daniel Roggen, Themis Palpanas, Elia Palme, Jürgen Scheffran, David Sumpter and Peter Wagner.

References

1. L. Backstrom, C. Dwork , J. Kleinberg, Wherefore Art Thou R3579X? Anonymized Social Networks, Hidden Patterns, and Structural Steganography. *Proc. 16th Int. World Wide Web Conference* (2007)

2. Apple confirms $1 bn data center, http://www.theregister.co.uk/2009/06/04/apple_1bn_north_carolina_data_center/. Accessed 10 Sept 2010

3. NSA plans massive, 65 MW, $2 bn data center in Utah, http://www.theregister.co.uk/2009/07/03/new_nsa_data_center/. Accessed 10 Sept 2010

4. Microsoft consumes Chicago data center, http://www.theregister.co.uk/2009/05/20/ascent_ch2_datacenter/. Accessed 10 Sept 2010

5. Google admits Scandinavian data center landing, http://www.theregister.co.uk/2009/03/05/google_finland_data_center/. Accessed 10 Sept 2010

6. Google pays $ 51.7 m for newspaper destruction metaphor, http://www.theregister.co.uk/2009/02/12/google_buys_defunct_paper_mill/. Accessed 10 Sept 2010

7. Intel sees future in Mega Data Center, http://www.theregister.co.uk/2009/02/18/the_intel_cloud/. Accessed 10 Sept 2010

8. D. Helbing, S. Balietti, From social simulation to integrative system design. Visioneer White Paper (2010), see http://www.visioneer.ethz.ch. Accessed 10 Sept 2010

9. C. Cattuto, W. Van den Broeck, A. Barrat, V. Colizza, J.-F. Pinton, A. Vespignani, Dynamics of person-to-person interactions from distributed RFID sensor networks. PLoS ONE **5**(7), e11596 (2010)

10. 123 People, http://www.123people.com. Accessed 10 Sept 2010

11. A. Mazlouimian, D. Helbing, Y.-H. Eom, S. Lozano, S. Fortunato, How citation boosts trigger scientific paradigm shifts. (in preparation) (2010)

12. D. Helbing, M. Treiber, N.J. Saam, Analytical investigation of innovation dynamics considering stochasticity in the evaluation of fitness. Phys. Rev. E **71**, 067101 (2005)

13. J. Lorenz, H. Rauhut, F. Schweitzer, D. Helbing, How social influence undermines the wisdom of crowds (2010) (Submitted)

14. S.E. Asch, Studies of independence and conformity: a minority of one against a unanimous majority. Psychol. Monogr. **70**(9) (1956)

15. R. Axelrod, *The Evolution of Cooperation*, (Basic Books, 1984), pp. 169–170

16. D. Helbing, W. Yu, K.-D. Opp, H. Rauhut, The emergence of homogeneous norms in heterogeneous populations. Am. J. Sociol. (2010) (submitted)

17. Oakland Crimespotting is an interactive map of crimes in Oakland and a tool for understanding crime in cities, http://oakland. crimespotting.org. Accessed 10 Sept 2010

18. Data, data everywhere. Econ. (25 February 2010)

19. Internet Reputation Management: neutralize negative publicity, http://www.internet-reputation-management.com/. Accessed 10 Sept 2010

20. Reputation Management Consultants, http://www.reputationmanagementconsultants.com/. Accessed 10 Sept 2010

21. Reputation Defender, http://www.reputationdefender.com/. Accessed 10 Sept 2010

22. Squidoo: internet reputation management, http://www.squidoo. com/internet-reputation-management. Accessed 10 Sept 2010

23. Micro Payment: professional payment provider, http://micropayment.de. Accessed 10 Sept 2010

24. Privatsphäre als Luxusgut, http://www.nzz.ch/blogs/nzz_blogs/betablog/privatsphaere_als_luxusgut_1.7266824.html. Accessed 10 Sept 2010

25. It seems it not so easy to clear ones name on-line, even when trying hard. This is specially true for traces left on social network Web sites, for which specific applications, such as http://suicidemachine.org/, have been created in order to accomplish this task. For a discussion based on a true story see http://ask.slashdot.org/story/09/12/10/2115238/Best-Way-To
-Clear-Your-Name-Online. Accessed 10 Sept 2010

26. F. Winter, H. Rauhut, D. Helbing, How norms can generate conflict. Jena Econ. Res. Papers (2009)

27. D. Helbing, A. Johansson, Cooperation, norms, and conflict: a unified approach. SFI Working Paper #09-09-040 (2009)

28. D. Helbing, W. Yu, The outbreak of cooperation among success-driven individuals under noisy conditions. Proc Nat Acad Sci USA (PNAS) **106**(8), 3680–3685 (2009)

29. D. Helbing, W. Yu, H. Rauhut, Self-organization and emergence in social systems. Modeling the coevolution of social environments and cooperative behavior. SFI Working Paper #09-07-026 (2009)

30. L. Lessig, Against Transparency, The New Republic, 9th Oct, 2009, http://www.tnr.com/article/books-and-arts/against-transparency. Accessed 10 Sept 2010

31. D. Helbing et al., Dynamic decision behavior and optimal guidance through information services: models and experiments, in *Human Behaviour and Traffic Networks*, ed. by M. Schreckenberg, R. Selten (Springer, Berlin, 2004), pp. 47–95

32. D. Helbing, M. Christen, Mit Rauschen und Reibung gegen finanzielle Blasen, submitted to Wirtschaftswoche (2010)

33. P. Bajaria, J. Yeo , Auction design and tacit collusion in FCC spectrum auctions. Inf. Econ. Policy **21**(2) 90–100 (2009)

34. C. Schultz, *Transparency and Tacit Collusion* (2001), http://www.econ.ku.dk/ansatte/phd/?pure=files%2F23184032%2F2001-04.pdf

35. B. Kluger, S.B. Wyatt, Preferencing, internalization of order flow, and tacit collusion: evidence from experiments. J. Financ. Quant. Anal. **37**(3), 449 (2002)

36. M. Mäs, A. Flache, D. Helbing, Individualization as driving force of clustering phenomena in humans. PLoS Comput. Biol. (2010) (in print)

37. Four million British identities are up for sale on the internet, http://technology.timesonline.co.uk/tol/news/tech_and_web/the_web/article6718560.ece. Accessed 10 Sept 2010

38. Symantec Internet Security Threat Report, http://www.symantec.com/business/theme.jsp?themeid=threatreport. Accessed 10 Sept 2010

39. RottenNeighbor.com was a website created to post information about neighbors and find information about new potential neighbors before moving. Launched in July 2007, it was discontinued in July 2009
40. D. Sally, I. Ron, H. Ursula, An EU Code of Ethics for Socio-Economic Research, *The Institute of Employment Studies*, 2004
41. E. Diener, R. Crandall, Ethics in social and behavioral research (University of Chicago Press, Chicago, 1978)
42. C. Frankfort-Nachmias, D. Nachmias, Research Methods in the Social Sciences (Worth Publishers, New York, 2008), Chap. 4: "Ethics in Social Research"
43. The British Psychological Society, Report of the Working Party on Conducting Research on the Internet, http://www.bps.org.uk/the-society/code-of-conduct/ (2007). Accessed 10 Sept 2010
44. Ethical Guidelines, Social Research Association (2003), http://www.the-sra.org.uk/ethical.htm. Accessed 10 Sept 2010
45. Statement of Ethical Practice for the British Sociological Association, BSA, the British Sociological Society (2002), http://www.britsoc.co.uk/equality/Statement+Ethical+Practice.htm. Accessed 10 Sept 2010
46. Recommendations of the Association of Internet Researchers (AoIR) Ethics Working Committee, http://www.aoir.org/reports/ethics.pdf. Accessed 10 Sept 2010
47. Google chief: only miscreants worry about net privacy, http://www.theregister.co.uk/2009/12/07/schmidt_on_privacy/. Accessed 10 Sept 2010
48. Google admits it accidentally gathered WiFi data, http://www.ft.com/cms/s/2/8a23b394-5fab-11df-a670-00144feab49a.html. Accessed 10 Sept 2010
49. Google to hand over intercepted data, http://www.ft.com/cms/s/2/db664044-6f43-11df-9f43-00144feabdc0.html. Accessed 10 Sept 2010
50. Lawyers Claim Google Wi-Fi Sniffing "Is Not an Accident", http://gizmodo.com/5554960/lawyers-claim-google-wi+fi-sniffing-is-not-an-accident. Accessed 10 Sept 2010

51. Wi-Fi Data Captured By Google Street View Cars Included Passwords, http://gizmodo.com/5567460/wi-fi-data-captured-by-google-street-view-cars-included-passwords. Accessed 10 Sept 2010

52. Security Focus, http://www.securityfocus.com. Accessed 10 Sept 2010

53. Did you watch porn? http://www.didyouwatchporn.com. Accessed 10 Sept 2010

54. What the Internet knows about you. This page checks your browser history and determines which of the 5000 most popular Internet websites you've recently visited, http://www.whattheinternet-knowsaboutyou.com. Accessed 10 Sept 2010

55. A. Janc, L. Olejnik , Feasibility and Real-World Implications of Web Browser History Detection, http://w2spconf.com/2010/papers/p26.pdf. Accessed 10 Sept 2010

56. P. Eckersley, How unique is your web browser? *Electronic Frontier Foundation* (2009)

57. Panopticlick: how unique and trackable is your browser? https://panopticlick.eff.org. Accessed 10 Sept 2010

58. The Electronic Frontier Foundation published a timeline of Facebook's privacy policy modifications over the years, http://www.eff.org/deeplinks/2010/04/facebook-timeline. Accessed 10 Sept 2010

59. Watchdog files complaint over Facebook 'privacy' settings, http://www.theregister.co.uk/2009/12/17/epic_facebook_privacy_complain/. Accessed 10 Sept 2010

60. Senator calls on FTC to tackle social-net privacy, http://news.cnet.com/8301-13577_3-20003415-36.html. Accessed 10 Sept 2010

61. EU warns on Facebook privacy, http://www.nytimes.com/2009/01/27/technology/27iht-facebook.4-417144.html. Accessed 10 Sept 2010

62. German minister warns Facebook over privacy rules, http://blog.foreignpolicy.com/posts/2010/04/05/german_minister_warns_facebook_over_privacy_rules. Accessed 10 Sept 2010

63. Details of 100 m Facebook users collected and published, http://www.bbc.co.uk/news/technology-10796584. Accessed 10 Sept 2010

64. G. Wondracek, T. Holz, E. Kirda, C Kruegel, A Practical Attack to De-Anonymize Social Network Users. Technical Report TR-iSecLab-0110-001

65. X. Su, T.M. Khoshgoftaar, A survey of collaborative filtering techniques. Adv. Artif. Intell., Article ID 421425 (2009), http://dx.doi.org/10.1155/2009/421425. Accessed 10 Sept 2010

66. E.J. Candes, T. Tao, The power of convex relaxation: near-optimal matrix completion. IEEE Trans. Inform. Theory **56**, 2053–2080 (2009)

67. M. Bezzi et al. (eds.), Privacy and Identity Management for Life. (Springer, Berlin, 2010) (5th IFIP Advances in Information and Communication Technology [Book 320])

68. G. Ziegler, C. Farkas, A. Lörincz, A framework for anonymous but accountable self-organizing communities. Inform. Softw. Technol. **48**, 726–744 (2006)

69. Government requests directed to Google and YouTube, http://www.google.com/governmentrequests/. Accessed 10 Sept 2010

70. Exclusive: Google, CIA Invest in 'Future' of Web Monitoring, http://www.wired.com/dangerroom/2010/07/exclusive-googlecia/. Accessed 10 Sept 2010

71. Apple's Worst Security Breach: 114,000 iPad Owners Exposed, http://gawker.com/5559346/apples-worst-security-breach-114000-ipad-owners-exposed. Accessed 10 Sept 2010

72. T-Mobile confirms biggest phone customer data breach, http://www.guardian.co.uk/uk/2009/nov/17/t-mobile-phonedata-privacy. Accessed 10 Sept 2010

73. Soziale Netzwerke verraten künftiges Käuferverhalten, http://www.tagesanzeiger.ch/digital/internet/Soziale-Netzwerke-verraten-kuenftiges-Kaeuferverhalten/story/19928880. Accessed 10 Sept 2010

74. WUALA, Backup. Store. Share. Access everywhere, http://www.wuala.com/. Accessed 10 Sept 2010

75. Tor: anonymity online, http://www.torproject.org/. Accessed 10 Sept 2010

76. Freenet, the free network, http://freenetproject.org/. Accessed 10 Sept 2010

77. Project Gaydar, http://www.boston.com/bostonglobe/ideas/articles/2009/09/20/project_gaydar_an_mit_experiment_raises_new_questions_about_online_privacy/. Accessed 10 Sept 2010

78. R. Agrawal, R. Srikant, Privacy-preserving data mining. *Proceedings of the 2000 ACM SIGMOD international conference on Management of data* (2000), pp. 439–450

79. P. Samarati, L. Sweeney, Generalizing Data to Provide Anonymity when Disclosing Information. *Proceedings of the Seventeenth ACM SIGACT-SIGMOD-SIGART Symposium on Principles of Database Systems* (ACM Press 1998)

80. C.C. Aggarwal, P.S. Yu, Privacy-Preserving Data Mining: Models and Algorithms (Springer, 2008), p. 530

81. M. Atzori, F. Bonchi, F. Giannotti, D. Pedreschi, Anonymity preserving pattern discovery. VLDB J. **17**(4), 703–727 (2006)

82. B.-C. Chen, D. Kifer, K. LeFevre, A. Machanavajjhala, Privacy-preserving data publishing (Survey). Found. Trends Databases **2**(1–2), 1–167 (2009)

83. F. Giannotti, D. Pedreschi, Mobility, Data Mining and Privacy: Geographic Knowledge Discovery (Springer, 2008), p. 410

84. A. Monreale, G. Andrienko, N. Andrienko, F. Giannotti, D. Pedreschi, S. Rinzivillo, S. Wrobel, Movement Data Anonymity through Generalization. Trans. Data Priv. **3**(2), 1–121 (2010), http://www.tdp.cat/issues/abs.a045a10.php. Accessed 10 Sept 2010

85. J. Krumm, A survey of computational location privacy. Pers. Ubiquitous Comput. **13**(6), 391–399 (2009)

86. D. Helbing, W. Yu, The future of social experimenting. PNAS **107**(12), 5265–5266 (2010)

87. The Future of Social Experimenting: The Full Story, http://www.soms.ethz.ch/research/socialexperimenting. Accessed 10 Sept 2010

88. Mechanical Turk is a market place for work, https://www.mturk.com/mturk/welcome. Accessed 10 Sept 2010

89. EFF: Forget cookies, your browser has fingerprints, http://www.computerworld.com/s/article/9176904/EFF_Forget_cookies_your_browser_has_fingerprints. Accessed 10 Sept 2010

90. R. Geambasu, T. Kohno, A. Levy, H.M. Levy, Vanish: Increasing Data Privacy with Self-Destructing Data. *Proceedings of the USENIX Security Symposium*, Montreal, Canada, August 2009

91. Vanish: self-destructing digital data, http://vanish.cs.washington.edu. Accessed 10 Sept 2010

92. QLectives (Quality Collectives), http://www.qlectives.eu. Accessed 10 Sept 2010

93. European privacy seals for IT products and IT-based services, https://www.european-privacy-seal.eu/. Accessed 10 Sept 2010

94. Ixquick: the world's most private search engine, http://ixquick.com/. Accessed 10 Sept 2010

95. Datenschutz fr iPhone-Apps, http://www.ethlife.ethz.ch/archive_articles/100930_MBusiness_Apps_sch/index. Accessed 10 Sept 2010

96. DIASPORA, The privacy aware, personally controlled, do-it-all, open source social network, http://www.joindiaspora.com/. Accessed 10 Sept 2010

97. EU Commission plans more harmonisation of data protection law, http://www.out-law.com/default.aspx?page=11228. Accessed 10 Sept 2010

98. K.P. Smith, N.A. Christakis, Social networks and health. Annu. Rev. Sociol. **34**, 405–429 (2008)

Further Reading

1. M. Michael, J.E. Moreira, D. Shiloach, R.W. Wisniewski, Scale-up x Scale-out: A Case Study using Nutch/Lucene. *Parallel and Distributed Processing Symposium, IEEE International*, 2007

2. L. A. Barroso, Hölzle, *The Datacenter as a Computer: An Introduction to the Design of Warehouse-Scale Machines*. (Morgan & Claypool Publishers, 2009)

3. A. Jacobs, The pathologies of big data. *ACM Queue*, **7**(6) (2009)

4. SSD Myths and Legends—"write endurance", http://www.storagesearch.com/ssdmyths-endurance.html. Accessed 10 Sept 2010

5. Digging into Data, http://www.diggingintodata.org/. Accessed 10 Sept 2010
6. Transparency is at the heart of this Government. Data.gov.uk is home to national & local data for free re-use, http://data.gov.uk.. Accessed 10 Sept 2010
7. Data.Gov Empowering people, http://www.data.gov. Accessed 10 Sept 2010
8. Dataverse Project: an open-source application for publishing, citing and discovering research data, http://thedata.org/. Accessed 10 Sept 2010
9. Apache WSIF: Web Service Invocation Framework, http://ws.apache.org/wsif/. Accessed 10 Sept 2010
10. ETH Financial Crisis Observatory, http://www.er.ethz.ch/fco/index. Accessed 10 Sept 2010
11. ICKN Galaxy Advisors, http://ickn.org. Accessed 10 Sept 2010
12. PostRank: intelligence from the social web, http://www.postrank.com/. Accessed 10 Sept 2010
13. M.M. Gaber, A. Zaslavsky, S. Krishnaswamy, Mining data streams: a review. ACM SIGMOD Rec. Arch. **34**(2), 18–26 (2005)
14. A. Bifet, R.K. August, Data Stream Mining: A Practical Approach. *The University of Waikato* (2009)
15. J. Leskovec, L. Backstrom, J. Kleinberg, Meme-tracking and the dynamics of the news cycle. *Proceedings of the 15th ACM SIGKDD international conference on Knowledge discovery and data mining* (2009), pp. 497–506
16. A. Narayanan, V. Shmatikov, Robust De-anonymization of Large Sparse Datasets. *IEEE Symposium on In Security and Privacy SP 2008.* IEEE Symposium (2008), pp. 111–125
17. R.Jones, R. Kumar, B. Pang, A.Tomkins, Vanity fair: privacy in query-log bundles. *CIKM '08: Proceeding of the 17th ACM conference on Information and knowledge management* (2008), pp. 853–862

12
What the Digital Revolution Means for Us

This chapter first appeared in Science Business on June 12, 2014, see http://www.sciencebusiness.net/news/76591/What-the-digital-revolution-means-for-us, and is reproduced here with minor stylistic improvements.

No country in the world is prepared for the digital era. We urgently need an Apollo-like programme and a Space Agency for information and communication technologies with a mission to develop the institutions and information infrastructures for the emerging digital society.

Never before were politicians, business leaders, and scientists more urgently needed to master the challenges ahead of us. We are in the middle of a third industrial revolution. While we see the symptoms, such as the financial and economic crisis, cybercrime and cyberwar, we haven't understood the implications well. But at the end of this socio-economic transformation, we will live in a digital society. This comes with breathtaking opportunities and challenges, such as occur only every 100 years.

12.1 Big Data: A magic Wand. But do we know How to Use it?

Let me start with Big Data. When the social messaging portal WhatsApp with its 450 million users was sold recently, it made

$ 19 billion, or almost $ 500 million per staff member. Big Data is fundamentally changing our world. It is becoming "the new oil of the twenty-first century", and we need to learn how to drill and refine it, that is, how to produce data and turn them into information, knowledge and wisdom.

The potential of Big Data spans all areas of society. It reaches from natural language processing over financial asset management, to smartly managing our cities and better balancing energy consumption and production, thereby saving energy. It allows for better protection of our environment, risk detection and reduction, and the discovery of opportunities, which would otherwise be missed. It will be possible to tailor medicine to patients, thereby increasing drug effectiveness while reducing side effects. Preventing diseases may become even more important than treating them.

Big Data applications are now spreading like wildfire. They enable personalized offers, services and products. Big Data open up entirely new possibilities for process optimization and allow one to identify unexpected interdependencies. They also imply great potential for evidence-based decision-making, but science will be crucial to ensure transparency, quality, and trust. Science will also be important to drive ethical ICT innovations and to avoid the pitfalls of Big Data applications. Therefore, science must become a fifth pillar of democracies, besides legislation, executive, jurisdiction, and the public media.

12.2 What Is the Next Big Thing After Big Data?

But we need to think a step ahead and realise that we are just at the beginning of a transformation process, which is about to change human history. The invention of the steam engine turned agricultural society (Economy 1.0) into industrial society (Economy 2.0), and wide-spread education turned it into service society (Economy 3.0). Now, the invention of computers, the Internet, the World Wide Web, and Social Media are transforming service societies into digital societies (Economy 4.0).

With computers reaching the level of human brainpower in about 10 years, the arrival of intelligent service robots, and the Big Data tsunami, 50 percent of jobs in the industrial and service sectors will probably be lost within the next 20 years. And most of our current ways of doing things will fundamentally change: the way we educate (MOOCS—Massively Open Online Courses—and personalized education), the way we do research (Big Data analytics), the way we move (self-driving Google cars) or transport goods (drones), the way we go shopping (take Amazon and eBay), the way we manufacture (3D printers), but also our health system (personalized medicine), and most likely politics (participation of citizens) and the entire economy as well (with the makers community, the emerging sharing economy, and prosumers, that is co-producing consumers). Financial business, which used to be the domain of banks, is increasingly replaced by algorithmic trading, Paypal, Bitcoin, Google Wallet, and so on. Moreover, the biggest share of the insurance business is now in financial products such as credit default swaps. Even wars may increasingly change from conventional wars to cyberwars.

Thus, how will the digital revolution transform our societies? First of all, the transition will be challenging. Today's world is struggling with financial instabilities, and in many areas of the world, we are faced with social and political unrest-sometimes framed as "Twitter revolutions". How can we handle this? Do we need more state power, based on armed police and mass surveillance? Could a giant supercomputer (or network or cloud of supercomputers), fuelled with massive amounts of data about human activities and almost everything, simulate our globe? Could a supercomputing infrastructure like this optimise and plan our world? Could it avoid the traps of particular interests, irrationality, and emotional decision-making? Could it find ways to overcome coordination and market failures, breakdowns of cooperation, and conflict? Could it take better decisions than we could do? And should it determine our actions through personalized recommendations and selective information that smartphones or other gadgets deliver to us?

To some or even many of us, this seems plausible, but this concept, known as "benevolent dictator" or "big government" cannot work. While the processing power doubles every 18 months, the amount of data doubles every year. Unfortunately, the complexity of networked systems is growing even faster (as the figure illustrates). In other words, attempts to optimize systems in a top-down way will be less and less effective—and can often not be done in real time. Paradoxically, as economic diversification and cultural evolution progress, a big government approach would increasingly fail to lead to good decisions. However, neither is simplifying our world by homogenization and standardization a solution—since it reduces innovation, societal resilience, and the happiness of people in general. Today, everyone already complains about over-regulation, and we can no longer pay for the expensive institutions needed for it. Most industrialized countries have reached historical heights in public debt levels in the order of 100 to 200 per cent of their annual productivity. Nobody knows how we should ever be able to pay for this—and for even more regulation.

But what alternatives are there? The logical answer is: distributed (self-)control, that is, bottom-up self-regulation, as envisioned by Adam Smith's paradigm of the invisible hand. While this vision was sometimes not working well in the past due to co-ordination and market failures, complexity theory tell us that it is actually feasible to create resilient social and economic order by means of self-organization, self-regulation, and self-governance. The work of Nobel prize winner Elinor Ostrom and others has demonstrated this. By "guided self-organization" we can let things

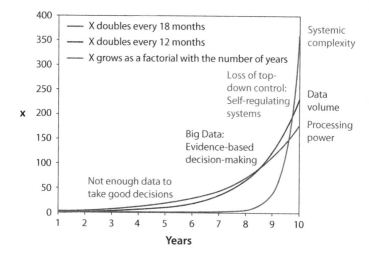

Fig. 12.1 Schematic illustration showing the exponential growth of process-ing power ("Moore's law"), the even faster exponential growth of the data volume stored, and the factorial growth of systemic complexity, as we go on networking the world, thereby creating new combinatorial possibilities. It becomes obvious that a decreasing share of data can be processed, which produces a "flashlight effect." While we focus on certain facts, we will ne-glect others. Therefore, we might pay too much attention to some issues, while forgetting to address others. For example, most people, including ex-perts, did not see the financial crisis coming, the Arab Spring, or the crisis in Ukraine, even though there must have been evidence for this. Moreover, as systemic complexity exceeds data volumes and processing power, a top-down optimization of systems is impossible in strongly variable, hardly predictable complex systems, as we have many of them. It's even unfeasible to perform a strict optimization of traffic light control in medium-sized and large cities. As a consequence, it is necessary to create a coordinated and desirable systemic outcome based on principles such as distributed control and self-organization

happen in a way that produces desirable outcomes in a flexible and efficient way. One should imagine this embedded in the framework of today's institutions and stakeholders, which will eventually learn to interfere in minimally invasive ways.

How will such self-regulation work? In a rapidly changing world, which is hard to predict and plan, we must create feedback loops that enable systems to flexibly adapt in real time to local conditions and needs. Now, 300 years after Adam Smith's historical vision, we can make it happen, fuelled by real-time data. For example, my research team has invented self-regulating traffic lights, which are driven by the traffic flows and can outperform the classical top-down control by a conventional traffic center. Can we transfer and extend this principle to socio-economic systems? Indeed, we are now developing mechanisms to overcome coordination and cooperation failures, conflicts, and other age-old problems. This can be done with suitably designed social media and sensor networks for real-time measurements, which will eventually weave a Planetary Nervous System. Hence, we can finally realise the dream of self-regulating systems, and there is now a rapidly increasing number of examples of them: Bitcoin, peer to peer lending, Google's self-driving car, Uber's limousine service, collaborative robot swarms, and social communities on the Web.

12.3   A New Kind of Economy is Born

A largely self-regulating society isn't utopia. In fact, a new kind of economy is already on its way. Social media connect people and, thereby, enable "collective intelligence." This paradigm is superior to the self-regarding optimization by the "homo economicus", the selfish decision-maker assumed in mainstream economics. While the bottom-up self-organization of the "homo economicus" can outperform top-down decision making in complex environments, highly competitive conditions can lead to coordination failures and poor outcomes (for example, "tragedies of the commons" such as environmental degradation). It has been theoretically and empirically shown, however, that a considerable fraction of people has other-regarding preferences—I will call this type "homo socialis." To understand the decisions of this type, a new economic thinking is needed compared to the purely selfish "homo economicus," which is the basis of the current mainstream economics. Considering so-called externalities, i.e. the impact of own decisions on others, enables self-regulation, which can overcome the above-mentioned coordination failures and "tragedies of the commons." Interestingly, suitable institutions such as certain social media—combined with suitable reputation systems—can promote other-regarding decision-making. The quick spreading of social media and reputation systems, in fact, indicates the emergence of a superior organizational principle, which creates collective intelligence by harvesting the value of diversity. Properly designed social media allow diverse knowledge and skills to come together, thereby unleashing creativity, social capital and productive value.

Hence, in accordance with the paradigm of distributed control and self-regulation, a participatory market society is on the rise. While the twentieth century was an era of democratization of consumption, with 3D printers and other new technologies, the twenty-first century can become an era of democratization of production. Next to todays companies, we see the emergence of an innovation ecosystem characterized by flexible, participatory forms of production, which I term "projects". Here, creative minds come together to realise joint project ideas. After completing a project, everyone looks for another one, and so on. Social media platforms such as Amazon Mechanical Turk make it possible to bring ideas and skilled workers together. As a consequence, this leads to a more direct participation of people in production processes (prosumers). Over time, there will be a much greater diversity of products, tailored to individual needs. Thus, while computers will increasingly replace our current types of routine and executive work, we will have an opportunity to replace these jobs by more creative activities. Production by large corporations will then be complemented by an innovation ecosystem made up of millions of projects. The huge range of smartphone apps that low-cost downloads from App stores have enabled, gives just a first idea of the unlimited possibilities for new projects. Open access data and the Web 2.0, Web 3.0, etc. will further accelerate this development.

12.4 The New Algebra of Prosperity and Leadership

The twenty-first century will be governed by fundamentally different principles than the twentieth century, and that's why we need to change our way of thinking about the world. To understand this, it is important to recognize the following facts and trends: information is ubiquitous and everywhere instantly available, such that borders dissolve. The "second machine age" comes with extreme speed. Most of our knowledge is outdated, and we can't learn quickly enough to fully understand the changing world without the help of smart devices such as "social information technologies." Many systems become more variable, less predictable, and less controllable. Their increased connectivity implies a higher complexity. The increase in data volumes means we are overloaded by data that ultimately needs to be converted into information and then into actionable knowledge. Furthermore, the more data we produce, the less likely can we keep secrets and the cheaper data will become. This means that we will make less profit on data, but more on algorithms that turn data sets into useful information and knowledge. In such a world, ideas will become more powerful, and ethics more important. Digitally literate people will be better informed than experts used to be, therefore, classical hierarchies will dissolve. Moreover, data can be replicated as often as we like. It's a virtually unlimited resource, which may help to overcome conflicts that scarce resources used to imply. However, services and products will be more individualized, personalized, and user-centric. Finally, what used to be science fiction may become reality. The countries first recognizing these new principles and turning them into their advantage will be leaders. Those failing to adapt to these trends in a timely manner will be in trouble. We may just have 20 years for this—a very short time considering that planning and building a road often takes 30 years or more.

12.5 What Does it Take to Master Our Future?

So far, no country in the world seems to be well prepared for the digital era. Therefore, we urgently need an Apollo-like program, and the equivalent of a Space Agency for ICT: an Innovation Alliance with the mission to develop the institutions and information infrastructures for the emerging digital society. This is crucial to master the challenges of the twenty-first century in a smart way and to unleash the full potential of information for our society. For illustration, it is helpful to recall the factors that enabled the success of the automobile age: the invention of cars and of systems of mass production; the construction of public roads, gas stations, and parking lots; the creation of driving schools and driving licenses; and last but not least, the establishment of traffic rules, traffic signs, speed controls, and traffic police. All of this required many billions each year. We invest a lot into the agricultural sector, the industrial sector, and also the service sector. But are we investing enough into the emerging digital sector?

What are the technological infrastructures and the legal, economic and societal institutions needed to make the digital age a big success? This question would set the agenda of the Innovation Alliance. A partial answer is already clear: we need trustworthy, transparent, open, and participatory ICT systems, which are compatible with our values. For example, it would make sense to establish the emergent *Internet of Things* as a *Citizen Web*. This would enable self-regulating systems through real-time measurements of the state of the world, which would be possible with a public information platform called the *"Planetary Nervous System."* It would also facilitate a real-time measurement and search engine: an open and participatory *"Google 2.0."*

To protect privacy, all data collected about individuals should be stored in a *Personal Data Purse* and, given informed consent, processed in a decentralized way by third-party *Trustable Information Brokers*, allowing everyone to control the use of their sensitive data. A *Micro-Payment System* would allow data providers, intellectual property right holders, and innovators to get rewards for their services. It would also encourage the exploration of new and timely intellectual property right paradigms (*"Innovation Accelerator"*). A pluralistic, *User-centric Reputation System* would promote responsible behavior in the virtual (and real) world. It would even enable the establishment of a new value exchange system called *"Qualified Money"*, which would overcome weaknesses of the current financial system by providing additional adaptability.

A *Global Participatory Platform* would empower everyone to contribute data, computer algorithms and related ratings, and to benefit from the contributions of others (either free of charge or for a fee). It would also enable the generation of *Social Capital* such as trust and cooperativeness, using next-generation User-controlled Social Media. A Job and Project Platform would support crowdsourcing, collaboration, and socio-economic co-creation. Altogether, this would build a quickly growing Information and Innovation Ecosystem, unleashing the potential of data for everyone: business, politics, science, and citizens alike.

We could also create a *Digital Mirror World* to explore the likely risks and opportunities of prospective decisions. Finally, *Interactive Virtual Worlds* would realise the full creative potential within different socio-economic settings and intellectual property right approaches. *Social Information Technologies* would help us to cope with the diversity resulting from this and to benefit from it. *Digital Literacy* and good education will be more important than ever. But with the emerging "Internet of Things" and participatory information platforms, we can unleash the power of information and turn the digital society into an opportunity for everyone. It just takes our will to establish the institutions required to make the digital age a great success. Are we ready for this?

13
Creating ("Making") a Planetary Nervous System as Citizen Web

This chapter first appeared as FuturICT blog on September 23, 2014,
see http://futurict.blogspot.de/2014/09/creating-making-planetary-
nervous.html, and is reproduced here with minor stylistic
improvements.

The goal of the Planetary Nervous System is to create an open, public, intelligent software layer on top of the "Internet of Things" as the basic information infrastructure for the emerging digital societies of the twenty-first century.

After the development of the computer, Internet, the World Wide Web, smartphones and social media, the evolution of our global information and communication systems will now be driven by the "Internet of Things" (IoT). Based on wirelessly connected sensors and actuators, it will connect "things" (such as machines, devices, gadgets, robots, sensors, and algorithms) with things, and things with people.

Already now, more things than people are connected to the Internet. In 10 years time, it is expected that something like 150 billion sensors will be connected to the IoT. Given such masses of sensors everywhere around us—sensors in our coffee machine, our fridge, our tooth brush, our shoes, our fire alarm etc.—the IoT could easily turn into a dystopian surveillance nightmare, if largely controlled by one company or by the state. For the IoT to be successful, people need to be able to trust the new information and communication system, and they need to be able to exert their right of informational self-determination, which also requires the possibility to protect privacy.

Most likely, the only way to establish such a trustable, privacy-respecting IoT is to build it as a Citizen Web. Citizens would deploy the sensors in their homes, gardens, and offices themselves, and they would decide themselves what sensor information to open up (i.e. decrypt), and for whom (and for how long). In other words, the citizens would be in control of the information streams. A software platform such as open Personal Data Store (openPDS) would allow everyone to manage the access to personal data produced by the IoT.

13.1   What are the Benefits of Having an "Internet of Things"?

1. One can perform real-time measurements of the (biological, technological, social and economic) world around us.

2. This information can be turned into (real-time) maps of our world and serve as compasses for decision-makers, enabling them to take better decisions and more effective actions, considering externalities

3. One can build self-organizing and self-regulating systems, based on real-time feedback and adaptation. Uses of these kinds will be enabled by a software layer that we call the "Planetary Nervous System" (PNS) or just "Nervous". It offers new possibilities that will allow humanity to overcome some long-standing problems (such as systemic instabilities or "tragedies of the commons" like environmental degradation, etc.), and to change the world to the better.

13.2 Basic Elements of the Planetary Nervous System

1. Sensor kits and smartphones, to measure the environment
2. Algorithms and filters to encrypt information or degrade it such that it is not sensitive anymore
3. Ad hoc network/mesh net (e.g. firechat) to enable direct communication between wirelessly communicating sensors
4. Server architecture to collect, manage and process data
5. A data analytics layer and possibly a search engine and Collective Intelligence/Cognitive Computing layer on top
6. An open Personal Data Store (such as openPDS) to empower users to exercise their right of informational self-determination
7. An app-store-like Global Participatory Platform (GPP) to share data, algorithms, and ratings
8. An editor allowing non-expert users to combine inputs and outputs in playful, creative ways
9. A multidimensional reputation and micro-payment system
10. A project platform to allow the Nervous community to coordinate and self-organize their activities and projects

We will build two variants of the Planetary Nervous System App for smart devices such as smartphones: Nervous and Nervous+. While Nervous would not save original sensor data, Nervous+ would potentially do so. Nervous is thought to be for users that are concerned about their personal data, while Nervous+ offers additional functionality for people who are happy to share data of all kinds. Hence, the users can choose the system they prefer.

Both Planetary Nervous System Apps would offer a rich Open Data stream accessible for everyone. They would build something like a "real-time data streaming Wikipedia", offering people and companies to build services and products on top. The PNS is hence an attempt to enable and catalyze new creative jobs in times where the digital revolution is expected to eliminate about 50 % of the conventional jobs of today.

13.3 Creating a Public Good, and Business and Non-Profit Opportunities for Everyone by Maximum Openness, Transparency, and Participation

The main goal of the PNS project is to create a public good, namely the basic information infrastructure for the emerging digital societies of the twenty-first century. Besides providing Open Data streams, the Planetary Nervous System may nevertheless offer some premium services to people and/or institutions, who pay for the services or have qualified to receive them for free (such as committed scientists or citizens). "Qualification" means contributions made to the components of the Planetary Nervous System, but also a responsible use of the information services. In this way, we want to reduce malicious uses of the powerful functionality of Nervous+ as much as possible.

The profits created by the PNS would be managed, for example, by a benefit corporation, which is committed to improving social and/or environmental conditions. The largest share of the profits should be used to promote the science, research and development promoting the PNS and services built on top of it. Profits created with inventions of the PNS shall also be used to support the PNS project.

As the PNS project wants to grow into a public good for everyone, the Planetary Nervous System project is committed to opening up its source codes, as much as this is not expected to create security issues or dangers to human rights. Depending on the competitive situation the PNS is in, the publication may be done with a delay (usually less than 2 years). To minimize delays we will create incentives for early sharing.

The goal of this strategy is to catalyze an open information and innovation ecosystem. Others will be able to use our codes (and other people's open source codes), modify them and share them back. The same will apply to data, Apps, and other contributions. In this way, the Nervous community will benefit maximally from contributions of other Nervous members, and everyone can build on functionality that has been created by others.

Contributions of volunteers will be acknowledged by mentioning the respective creators by name (if they don't prefer to stay anonymous or pseudonymous). In addition, contributions will be rewarded by ratings, reputational values, or scores, which may be later used to get access to premium services. These would include larger query or data volumes ("power users") or an earlier access to codes that will be publicly released with a delay, or further benefits. The PNS project may also hand out medals or prizes for outstanding contributions, or highlight them in social or public media.

13.4   The Role of Citizen Science

For the Planetary Nervous System to be successful, it is crucial to develop a large community of users, but the underlying logic of sharing, bottom-up involvement and informational self-determination demands that everyone is encouraged to contribute to the creation of the system itself. The system would hence be built similar to Wikipedia or OpenStreetMap. In fact, the success of OpenStreetMap is based on the contributions of 1.5 million volunteers worldwide.

This is, why the Nervous project wants to engage with Citizen Science, to grow the Planetary Nervous System as a Citizen Web. As basis of citizen engagement, the Nervous Team will provide (a) kits containing sets of sensors and actuators (e.g. a basic kit, and several extension kits) and (b) a GPP portal, where people can download (and upload) algorithms ("Apps"), which will run on the sensors and thereby produce certain kinds of functionalities.

The Citizen Science community will be engaged in certain measurement tasks (e.g. "measure the noise distribution in your city as a function of time", or "measure data enabling weather predictions"). It will also be engaged to come up with innovative ways to use sensor data and turn them into outputs (i.e. to produce new codes or modify existing ones, thereby creating new Apps). For this, the PNS team will provide tools (such as an editor), allowing non-expert users to transform inputs into outputs in playful, creative ways. Playfulness, fun and reputation are hence offered in exchange for contributing to the development and spreading of the PNS. As a result, we will get new measurement procedures for science, and adaptive feedback processes to create self-regulating systems.

Printed in Great Britain
by Amazon

61247115R00119